ITIL® Exam Prep

Questions, Answers & Explanations

ITIL® Exam Prep

Questions, Answers & Explanations

Christopher Scordo, PMP, ITIL

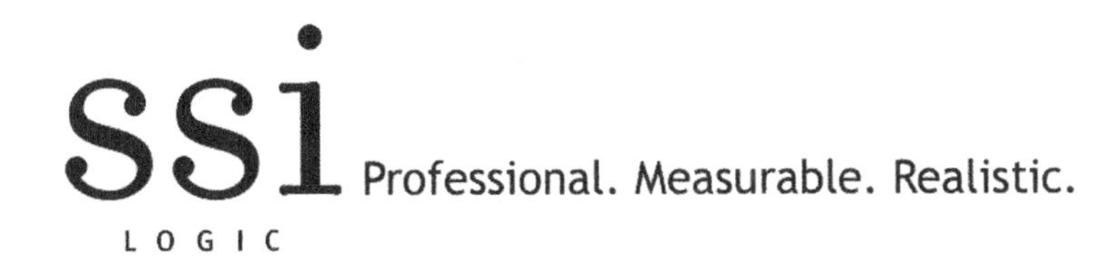

Revised Edition 2012.

Published by SSI Logic

ISBN-10: 0982576811
ISBN-13: 978-0-9825768-1-6

All inquiries should be addressed via email to:
support@ssilogic.com

or by mail post to:
SSI Solutions, INC
340 S Lemon Ave #9038
Walnut, CA 91789

Table of Contents

Additional Resources

INTRODUCTION

Welcome

Thank you for selecting SSI Logic's *ITIL® Exam Prep – Questions, Answers, and Explanations* for your ITIL Foundation study needs. The goal of this book is to provide full mock exams and practice tests which allow you to become comfortable with the pace, subject matter, and difficulty of the ITIL Foundation exam.

In this, the 2012 Edition of the text, question explanations have been updated to reinforce ITIL concepts; content improvements have been applied to reflect the "ITIL 2011 Edition" updates applied to the exam on **January 1, 2012**.

The content in this book is designed to optimize the time you spend studying in multiple ways.

1. Practice exams in this book reflect the actual length and subject matter of the ITIL Foundation exam and are designed to be completed in one hour; allowing you to balance your time between practice tests and offline study.

2. Passing score requirements in this book are slightly higher than the real exam; allowing you to naturally adjust to a higher test score requirement.

3. Practice exams included in this book cover the entire scope of the ITIL Foundation syllabus and ITIL core volumes, while shorter quizzes focus only on specific ITIL syllabus knowledge areas.

The practice exam content in this book is structured into two general types of exam preparation:

- Mock exams which allow you to test your knowledge across multiple iterations of the ITIL Foundation exam; designed to be completed within one hour.

- Knowledge Area Quizzes, which reflect brief practice tests focused on specific ITIL Foundation syllabus areas; designed to be completed in 10 to 20 minutes.

We wish you the best of luck in your pursuit of the ITIL Foundation certificate.

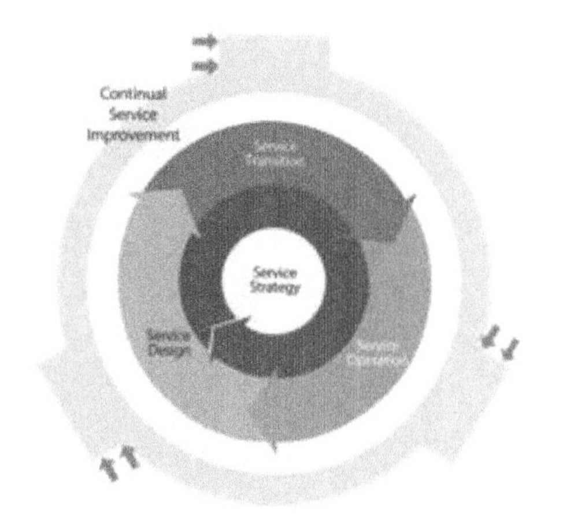

Diagram: ITIL Service Lifecycle –

ITIL® Foundation Exam Overview

The ITIL practice questions in this book reflect the ITIL Foundation exam version based on the ITIL core volumes.

About the IT Infrastructure Library (ITIL) Foundation Certification

Like all ITIL certifications, the *ITIL Foundation certificate in IT Service Management* (ITIL Foundation certification), is managed by the ITIL Certification Management Board (ICMB). The ICMB is composed of various stakeholders in the global ITIL community; including the UK Office of Government Commerce (OGC), the APM Group (APMG), various Examination Institutes (EIs), and other interested parties. Since the early 2000s, the ITIL Foundation certification has grown to become one of the most sought after internationally recognized management credentials available; consistently ranking in the top 5 industry certifications across IT career surveys.

The overall ITIL qualification scheme relies on the ITIL Foundation certificate as the first core requirement prior to attaining advanced ITIL certifications. Further ITIL certifications are available at the Intermediate, Expert, and Master levels. Attaining the ITIL Foundation certificate is required prior to acquisition of other ITIL certifications.

ITIL Foundation Exam Details

The ITIL Foundation exam is designed to objectively assess and measure knowledge reflecting the ITIL service lifecycle approach to IT management. Concepts covered in the Foundation exam are directly derived from the five ITIL core volumes, with subject matter categorized among the ITIL Foundation syllabus knowledge areas.

The actual exam is offered in a computer based testing (CBT) environment through a number of Examination Institutes (EIs). EIs which offer the exam include EXIN, BCS-ISEB, Loyalist Certification Services (LCS), CSME, and APMG. All EIs offer ITIL Foundation exams which reflect the predefined Foundation syllabus; with little or no variation between the exams they administer.

A summary of the exam structure and passing requirements are as follows:

- There are 40 total multiple choice questions which make up the ITIL Foundation exam
- Individuals have 60 minutes to complete the exam
- Individuals must score 65% or higher to pass the exam (26 of 40 questions)

The ITIL Foundation Syllabus

The ITIL Foundation syllabus reflects the subject matter taxonomy covered in the exam, and is comprised of concepts from each of the five ITIL core volumes. In short, the exam taker is expected to demonstrate comprehension of the following learning objectives; or "knowledge areas":

- Service Management as a Practice
- The ITIL Service Lifecycle
- Generic Concepts and Definitions
- Key Principles and Models
- Selected Processes
- Selected Functions
- Selected Roles
- Technology and Architecture
- Competence and Training

The ITIL practice exam content in this book include practice quizzes targeted at each of the ITIL Foundation syllabus knowledge areas.

The knowledge areas of the ITIL Foundation syllabus reflect the concepts presented within the **five ITIL core volumes**. ITIL's five core volumes, also known as the ITIL Library, are:

1. Service Strategy
2. Service Design
3. Service Transition
4. Service Operation
5. Continual Service Improvement

The ITIL Qualification Scheme

The hierarchy of available ITIL certifications is known as the *ITIL qualification scheme.* As mentioned earlier, the entry level for the scheme is the **ITIL Foundation** certification. Once the ITIL Foundation Level is successfully completed, the candidate becomes eligible to take further ITIL certifications within the scheme.

The **ITIL Intermediate Level** provides two specific streams of education. One stream is focused on Service Lifecycle "Management"; while the other stream is focused on Service Capability "Process". Each stream is composed of multiple modules, and each module focuses on a different area of the ITIL spectrum. The ITIL Intermediate level is widely considered to lie at the core of the ITIL qualification scheme.

The next level of the ITIL qualification scheme is the **ITIL Expert Level**. The ITIL Expert Level is awarded to candidates who have achieved multiple certificates reflecting ITIL and earlier ITIL schemes; and also by achieving related qualifications which reinforce ITIL and the core certification program.

The highest level of the ITIL qualification scheme is the **ITIL Master Level**. This is the final qualification available within the ITIL program and is reserved for those professionals who have demonstrated evidence of their ability to successfully implement ITIL framework concepts and IT Service Management Best Practices within industry.

PRACTICE EXAMS AND QUIZZES

ITIL Foundation
Practice Exam 1
Practice Questions

Test Name: ITIL Foundation Practice Exam 1
Total Questions: 40
Correct Answers Needed to Pass: 30 (75.00%)
Time Allowed: 60 Minutes

Test Description

This is a cumulative ITIL Foundation test which can be used as a baseline for initial performance. This practice test includes questions from all ITIL question categories.

Test Questions

1. What concept below reflects "fitness for purpose" and represents the characteristics of a service which a customer gets to achieve desired outcomes?

   A. Warranty

   B. Utility

   C. Resources

   D. Service Management

2. Which of the following ITIL Core volumes places and prioritizes improvement programs according to strategic objectives?

   A. Service Strategy

   B. Continual Service Improvement

   C. Service Transition

   D. Service Operation

3. Which of the statements below are true regarding service assets?

   I. The justification to maintain and upgrade service assets is increased as the demand for those service assets increase
   II. Costs incurred by accommodating the demand for services are recovered from the customer through agreed terms and conditions

   A. I and II

   B. I

   C. II

   D. Both statements are false

4. Which of the following are valid objectives of the ITIL Service Desk function?

    I. To quickly restore normal service to users after an interruption
    II. To provide first-line support to users
    III. To manage the resolution of incidents
    IV. To escalate requests which cannot be resolved by first-line support

    A. I, II

    B. I, II, IV

    C. II, III, IV

    D. All of these responses / All of the above

5. What role acts as the primary point of contact to customers for all service related questions and issues, ensures that customer delivery and support requirements are met, and may identify opportunities for service improvements which result in raised RFCs?

    A. Service Owner

    B. Process Owner

    C. Change Manager

    D. Service Level Manager

6. Under ITIL, what are the 4 "P"s which facilitate effective Service Management?

    A. People, Processes, Products, Partners

    B. Profit, Procedure, Products, Potential

    C. People, Procedure, Profit, Planning

    D. People, Products, Profit, Performance

7. What generic Service Management concept(s) allows defined processes such as an organization's Incident Lifecycle and Change Model to be pre-defined and controlled, resulting in the automatic management of items such as escalation paths and alerting?

    A. Service Assets

    B. Workflow or process engines

    C. Incident workarounds

D. The Deming Cycle

8. Which of the Service Strategy statements below are true?

I. Both Resources and Capabilities are types of assets
II. Capabilities are easier for an organization to acquire than Resources

A. I and II

B. I

C. II

D. Both statements are false

9. What term best represents the concept of outcomes desired by customers, in combination with the services available to facilitate those outcomes?

A. Service Portfolio

B. Service Catalogue

C. Market Space

D. Market Value

10. What aspect of Service Design is concerned with designing the calculation methods and metrics of services?

A. Technology archictectural design

B. Process Design

C. Measurement Design

D. Service Portfolio Management

11. What type of Service Desk is typically dispersed geographically, provides 24 hour support at a relatively low cost, and may need to provide safeguards which address different cultures?

A. Local Service Desk

B. Centralized Service Desk

C. Virtual Service Desk

D. Follow the Sun

12. A large law firm is evaluating a new email service offered by a managed service provider. The law firm is determining if the service meets a specific functional requirement. The ability of the email service to do so is best described by which of the following terms?

A. Warranty

B. Value

C. Utility

D. Capacity

13. What statement below best describes the concept of "Service Management" in ITIL?

A. A set of specialized organizational capabilities for providing value to customers in the form of services.

B. A logical concept referring to people and automated measures that execute a defined process, an activity, or a combination thereof.

C. A means of delivering value to customers by facilitating outcomes they want to achieve, without the ownership of specific costs and risks.

D. A team, unit, or person that performs tasks related to a specific process.

14. What types of context-sensitive tools execute automated scripts which assist with diagnosing incidents in a short amount of time?

A. Remote control tools

B. Discovery, Deployment, and Licensing Technologies

C. Integrated Configuration Management System (CMS)

D. Diagnostic utilities

15. You are responsible for documenting the required level of service and performance targets with your organization's internal Application Support group so that the expected level of service can be delivered to the customer. What is this internal, underpinning document referred to as?

A. Operational Level Agreement (OLA)

B. Availability Management Plan

C. Service Level Agreement (SLA)

D. Service Catalogue

16. Which of the following statements regarding the Service Portfolio are true?

I. The Service Portfolio should form part of a Service Knowledge Management System (SKMS)

II. The Service Portfolio should be registered as a document within an organization's Configuration Management System (CMS)

A. I and II

B. I

C. II

D. None of the above are true.

17. Which of the following is not a key metric associated with Service Level Management?

A. SLA Targets Missed

B. SLA Breaches in Underpinning Contracts

C. Number of Service Interruptions

D. Customer Satisfaction of SLA Achievements

18. An item such as a service component or an asset of the organization which is under the control of Configuration Management is known as what?

A. Configuration Item (CI)

B. Configuration Management System (CMS)

C. Variant

D. Resource

19. What role is responsible for identifying and documenting the value of services within an organization and provides cost information to Service Portfolio Management?

A. IT Financial Manager

B. Service Level Manager

C. Demand Manager

D. Product Manager

20. What statement below is true regarding a Function?

A. A function must be performed by multiple personnel, representing an organizational unit

B. A single person must focus on a single function only

C. Functional hierarchies decrease cross-functional coordination within an organization

D. A function optimizes work methods by focusing on specific outcomes

21. What Service Design process ensures agreed levels of availability from the SLA is met, and maintains an Availability Plan which reflects the current and future needs of the organization?

A. Service Level Management

B. Capacity Management

C. Availability Management

D. Service Catalogue Management

22. What Service Operation principle focuses on the ability to respond to changes without impacting other services?

A. Responsiveness

B. Quality of Service

C. Cost of Service

D. Stability

23. Which of the following items do not represent a valid term for the structure of a Service Desk?

A. Centralized Service Desk

B. Production Service Desk

C. Local Service Desk

D. Virtual Service Desk

24. Which of the following ITIL Core volumes relies on practices and methods from Quality Management, Change Management, and the enhancement of capabilities?

A. Service Design

B. Continual Service Improvement

C. Service Operation

D. Service Transition

25. What item is responsible for storing only those services which are currently active and being offered by the organization?

A. Service Catalogue

B. Service Pipeline

C. Service Specification

D. Service Portfolio

26. What role is responsible for ensuring IT capacity is adequate for the delivery of services, produces and maintains a Capacity Plan, and balances capacity with demand?

A. Capacity Manager

B. Availability Manager

C. Service Level Manager

D. Demand Manager

27. In which of the following areas can Service Management be positively impacted by automation?

I. Design and modeling
II. Service Catalogue
III. Classification and routing
IV. Optimization

A. I and II

B. II and III

C. I, III and IV

D. All of these responses / All of the above

28. A Service Desk structure which utilizes a number of sites, located in different time zones, in an effort to provide 24 hour support is referred to as:

A. Full coverage support

B. Full service support

C. Follow the sun support

D. Follow the moon support

29. What concept allows management to better understand a service's quality requirements, and presents both the associated costs and expected benefits?

A. Business Case

B. Technical Service Catalogue

C. Risk Analysis

D. Service Portfolio

30. Under Service Design, what process is used to manage Underpinning Contracts with vendors through their lifecycle; and assures that those Underpinning Contracts are aligned with targets specified in customer SLAs?

A. Supplier Management

B. Capacity Management

C. Service Level Management

D. Information Security Management

31. Which of the following is not a Core volume of the ITIL Library?

A. Service Strategy

B. Service Transition

C. Continual Service Improvement

D. Service Catalogue Management

32. In ITIL, the activity of planning and regulating a process, with the objective of performing a process in an effective, efficient and consistent manner is known as what?

A. The Process Model

B. Process Change

C. Process Control

D. Process Policy

33. What role is responsible for formally authorizing changes, and may delegate this responsibility to another role based on pre-defined parameters such as risk and cost?

A. Product Manager

B. Change Authority

C. Change Manager

D. Change Advisory Board

34. Which of the following items is not a recognized Change Type under Service Transition?

A. Normal

B. Standard (Pre-authorized)

C. Emergency

D. Planned

35. Which of the following items below would least likely be categorized as a Service Request?

A. A request for a password re-set to allow a user to log in to a time tracking program

B. A request for a user manual for the corporate extranet

C. A request to fix a user's network connection which failed after an application caused an error

D. A request for a database extract from an information system

36. What role is responsible for owning, maintaining, and protecting the Known Error Database?

A. Incident Manager

B. Problem Manager

C. Alert Manager

D. Event Manager

37. Under Continual Service Improvement (CSI), what type of metrics are computed from component metrics?

A. Technology Metrics

B. Process Metrics

C. Baseline Metrics

D. Service Metrics

38. What Service Strategy process quantifies the financial value of IT Services, assets which underlie the provisioning of services, and the qualifications of operational forecasting?

A. Demand Management

B. Financial Management

C. Strategy Generation

D. Service Portfolio Management

39. The ability of an organization to transform resources in to services which can offer value to customers is a critical concept for which of the following?

A. Process Model

B. Service Management

C. Role

D. Good Practice

40. During a client discussion, you are asked to summarize the meaning of an Incident. Which of the items below most appropriately defines an Incident?

A. The unknown root cause of one or more disruptions to Service

B. Any event which may lead to the disruption, or decreased quality, of a Service

C. A disruption for which the root cause is known

D. A support request not involving a failure in the IT infrastructure.

ITIL Foundation
Practice Exam 1
Answer Key and Explanations

1. B - Utility reflects "fitness for purpose" and is what the customer gets to achieve desired outcomes. (Service Management) - Generic Concepts and Definitions [ITIL - Generic Concepts and Definitions]

2. B - Continual Service Improvement accomodates learning and improvement, and prioritizes improvement programs and projects in accordance with an organization's strategic objectives. (Service Lifecycle) [ITIL - Service Lifecycle]

3. A - Both statements above are true. As more demand is generated for services, there is more reason to keep those services maintained. The cost of offering a service to a customer is recovered from the customer.(ITIL Service Strategy) - Service Management as a practice [ITIL - Service Management as a Practice]

4. D - All of the items listed are valid objectives of the Service Desk Function. (Service Operation) - Selected Functions [ITIL - Selected Functions]

5. A - The Service Owner is responsible for the activities listed above; In addition, the Service Owner will communicate with Process Owners over the course of the Service Management lifecycle. (Service Management / Service Lifecycle) - Selected Roles [ITIL - Selected Roles]

6. A - The 4 "P"s of Service Management, as defined by ITIL, are People, Processes, Products, Partners. (Service Design) - Key Principles and Models [ITIL - Key Principles and Models]

7. B - Workflows and/or process engines allow processes to be pre-defined and controlled, resulting in automatic management of the items described. (Service Management) - Technology and Architecture [ITIL - Technology and Architecture]

8. B - Resources are easier for an organization to acquire than Capabilities; Capabilities relate to the management, processes, and knowledge within the organization. (Service Strategy) - Generic Concepts and Definitions [ITIL - Generic Concepts and Definitions]

9. C - Market Space is represented by the desired outcomes of customers, along with the services available to support those outcomes. (Service Strategy) -

Selected Processes [ITIL - Selected Processes]

10. C - In Service Design, Measurement Design is concerned with with designing the calculation methods and metrics of services. (Service Design) - Key Principles and Models [ITIL - Key Principles and Models]

11. D - The characteristics above describe a Follow the Sun approach. (Service Operation) - Selected Functions [ITIL - Selected Functions]

12. C - Utility is the functionality offered by a service to meet a particular need. Utility is often defined as "what the service does." [ITIL - Generic Concepts and Definitions]

13. A - ITIL describes Service Management as "A set of specialized organizational capabilities for providing value to customers in the form of services.". (ITIL Service Strategy) - Service Management as a practice [ITIL - Service Management as a Practice]

14. D - Diagnostic utilities use automated scripts to diagnose incidents earlier, and are context sensitive regarding the application. (Service Management) - Technology and Architecture [ITIL - Technology and Architecture]

15. A - An OLA would document the level of service and performance target expected of internal departments within the organization. (Service Design) - Generic Concepts and Definitions [ITIL - Generic Concepts and Definitions]

16. A - The Service Portfolio should be part of an overarching SKMS, and should also be registered as a document in the CMS. (Service Design) - Key Principles and Models [ITIL - Key Principles and Models]

17. C - The number of Incidents due to Capacity Shortages would not be tracked by Service Level Management; rather it would be tracked by Availability Management. (Service Design) - Selected Processes

18. 18.) A - ? A Configuration Item (CI) includes any item which is, or will be, under the control of Confguration Management. (Service Transition) - Generic Concepts and Definitions [ITIL - Generic Concepts and Definitions]

19. A - The IT Financial Manager is responsible for identifying and documenting the value of services within an organization and provides cost information to Service Portfolio Management. (Service Strategy) - Selected Roles [ITIL - Selected Roles]

20. D - A function can be performed by as little as one person within an organization. In smaller organizations, single persons may be multi-functional. Functional hierarchies increase cross-functional coordination, eliminating silos within an organization.(ITIL Service Design) - Service Management as a practice [ITIL - Service Management as a Practice]

21. C - As the name suggests, Availability Management ensures agreed levels of availability from the SLA is met, and maintains an Availability Plan which reflects the current and future needs of the organization. (Service Design) - Selected Processes [ITIL - Selected Processes]

22. A - "Responsiveness" focuses on the ability to respond to changes without impacting other services. (Service Operation) - Key Principles and Models [ITIL - Key Principles and Models]

23. B - The common terms used to define options for structuring Service Desks are Centralized Service Desk, Local Service Desk, Virtual Service Desk, and the Follow the Sun approach. (Service Operation) - Selected Functions [ITIL - Selected Functions]

24. B - Continual Service Improvement integrates the practices and methods from Quality Management, Change Management, and the improvement of capabilities. (Service Lifecycle) [ITIL - Service Lifecycle]

25. A - The Service Catalogue consists of services currently being offered. The Service Pipeline consists entirely of services in development. The Service Portfolio consists of services in development, services currently being offered, and services which have been retired. (Service Lifecycle) - Generic Concepts and Definitions [ITIL - Generic Concepts and Definitions]

26. A - The Capacity Manager performs the activities above, and also configures monitoring of capacity via different types of performance reporting. (Service Design) - Selected Roles [ITIL - Selected Roles]

27. D - Service Management can benefit from automation in all of the areas listed; in addtion, pattern recogniation and analysis, and detection and monitoring may also be improved. (Service Management) -

Technology and Architecture [ITIL - Technology and Architecture]

28. C - As the term suggests, "Follow the sun support" aims to provide users with around the clock support. (Service Operation) - Selected Functions [ITIL - Selected Functions]

29. A - A Business Case presents management with a service's quality requirements and associated delivery costs, in addition to models which outline what a service is expected to achieve. (Service Lifecycle) - Generic Concepts and Definitions [ITIL - Generic Concepts and Definitions]

30. A - Underpinning Contracts are agreements made with 3rd party suppliers/vendors, and must be in alignment with customer SLAs. Supplier Management ensures this. (Service Design) - Selected Processes [ITIL - Selected Processes]

31. D - The Core volumes of ITIL are Service Strategy, Service Design, Service Transition, Service Operation, and Continual Service Improvement (Service Lifecycle) [ITIL - Service Lifecycle]

32. C - Process Control plans and regulates processes so that they are performed effectively, efficiently, and consistently. (ITIL Service Design) - Service Management as a practice [ITIL - Service Management as a Practice]

33. B - The Change Authority (which is a given role, person, or group) is responsible for the activities listed above. (Service Transition) - Selected Roles [ITIL - Selected Roles]

34. D - The three Change Types defined by ITIL are Normal (not pre-approved), Standard (Pre-authorized), and Emergency. (Service Transition) - Generic Concepts and Definitions [ITIL - Generic Concepts and Definitions]

35. C - As Service Requests are classified as incidents which do not reflect a failure in the IT infrastructure, the user who's application caused an error would be the correct response. (Service Operation) - Selected Functions [ITIL - Selected Functions]

36. B - The Problem Manager is responsible for the Known Error Database, and initiates the formal closure of all Problem records. (Service Operation) - Selected Roles [ITIL - Selected Roles]

37. D - Service Metrics are computed from component metrics, and are the result of the end-to-end service. (Continual Service Improvement) -

Key Principles and Models [ITIL - Key Principles and Models]

38. B - Financial Management is the Service Strategy process responsible for quantifying the financial value of IT services, assets, and qualificiations of operational forecasting. (Service Strategy) - Selected Processes [ITIL - Selected Processes]

39. B - The statement describes a core concept of Service Management in ITIL. (ITIL Service Strategy) - Service Management as a practice [ITIL - Service Management as a Practice]

40. B - quality to a Service. A support request not resulting from a failure in the IT infrastructure is a type of Incident (Service Request), but is not broad enough. (Service Operation) - Generic Concepts and Definitions [ITIL - Generic Concepts and Definitions]

Knowledge Area Quiz: Service Management as a Practice Practice Questions

Test Name: Knowledge Area Quiz: Service Management as a Practice
Total Questions: 10
Correct Answers Needed to Pass: 7 (70.00%)
Time Allowed: 10 Minutes

Test Description

This practice test specifically targets the subject area of Service Management as a Practice.

Test Questions

1. What is the proper sequence of the ITIL concepts listed below?

I. Good Practice
II. Best Practice
III. Evolution in to Commodity, Generally accepted principles, Perceived Wisdom, or Regulatory requirements

A. II occurs first, I occurs second, III occurs last

B. I occurs first, III occurs second, II occurs last

C. B occurs first, C occurs second, A occurs last

D. A occurs first, B occurs second, C occurs last

2. What statement below best describes the concept of a "Service" in ITIL?

A. A set of specialized organizational capabilities for providing value to customers in the form of services.

B. A logical concept referring to people and automated measures that execute a defined process, an activity, or a combination thereof.

C. A means of delivering value to customers by facilitating outcomes they want to achieve, without the ownership of specific costs and risks.

D. A team, unit, or person that performs tasks related to a specific process.

3. Which of the statements below are true about the ITIL concept of "Good Practice"?

I. Good Practice represents Best Practices which have been

commonly accepted and applied throughout the industry

II. Good Practice is often referred to as the "most appropriate" and is considered to be complete, with no gaps

III. Good Practice reflects an approach to an undertaking which has not yet been proven to be successful

A. I

B. I and II

C. III

D. All of these responses / All of the above

4. Which of the statements below describe the value offered to customers through the use of a Service?

I. Performance of associated tasks are enhanced

II. The probability of desired outcomes on behalf of the customer is increased

III. Customer risk is increased, along with a higher potential for reward

IV. The effect of constraints are reduced

A. I, III, IV

B. II, III, IV

C. I, II, IV

D. All of these responses / All of the above

5. What statement below best describes the concept of a "Role" in ITIL?

A. A set of specialized organizational capabilities for providing value to customers in the form of services.

B. A team, unit, or person that performs tasks related to a specific process.

C. A logical concept referring to people and automated measures that execute a defined process, an activity, or a combination thereof.

D. A means of delivering value to customers by facilitating outcomes they want to achieve, without the ownership of specific costs and risks.

6. The technical support group within an organization is expected to assist with corporate software releases every quarter through specific Release Management activities, as well as ensure the network is consistently accessible via Availability Management activities. What statement below is true?

A. Because they are a single department, the technical support group is playing a single role regardless of the activities they perform.

B. The technical support group exists in a functional silo.

C. The technical support group is performing functions outside of its authority.

D. The technical support group is playing multiple roles by performing these activities.

7. Which of the following statements best describes the relationship between procedures and work instructions?

A. A procedure describes who should carry out logically related activities, while work instructions define how activities in a procedure should be carried out at a highly detailed level.

B. A work instruction describes who should carry out logically related activities, while procedures define how activities in a work instruction should be carried out.

C. A work instruction may include activities and stages from different processes, while a procedure only focuses on a single activity within a work instruction.

D. A work instruction only focuses on who must complete a given unit of work, while a procedure only focuses on how the work will be performed.

8. Which of the following items fall under the area of Process Enablers in the Process Model?

I. Process Owner
II. Process Resources
III. Process Capabilities
IV. Process Policy

A. I, IV

B. II, III

C. I, II, IV

D. II, III, IV

9. An effective approach to an undertaking which has already proven to be successful, but has not yet become common industry practice, is known as what?

A. Good Practice

B. Generally Accepted Principles

C. Best Practice

D. Perceived Wisdom

10. Within a process, the description of logically related activities, along with who should carry out the activities (i.e. "who does what"), is represented by which term?

A. Work Instruction

B. Best Practice

C. Function

D. Procedure

Knowledge Area Quiz: Service Management as a Practice Answer Key and Explanations

1. A - Under ITIL, Best Practice adoption or development of Best Practice occurs first, over time the industry will adopt the Best Practice, turning it in to a Good Practice, which will then undergo Evolution in to Commodity, Generally accepted principles, Perceived Wisdom, or Regulatory requirements. (General ITIL) - Service Management as a practice [ITIL - Service Management as a Practice]

2. C - ITIL describes a Service as "A means of delivering value to customers by facilitating outcomes they want to achieve, without the ownership of specific costs and risks.". (ITIL Service Strategy) - Service Management as a practice [ITIL - Service Management as a Practice]

3. B - Good Practice has already evolved from Best Practice, and is considered to be proven and successful. (General ITIL) - Service Management as a practice [ITIL - Service Management as a Practice]

4. C - ITIL considers the value offered to customers through a Service to be all of the items above, except for increased risk. A Service should lower the risk to the customer for completion of associated tasks. (ITIL Service Strategy) - Service Management as a practice [ITIL - Service Management as a Practice]

5. B - ITIL describes a Role as "A team, unit, or person that performs tasks related to a specific process.".(ITIL Service Strategy) - Service Management as a practice [ITIL - Service Management as a Practice]

6. D - A single group within an organization may play several roles, as demonstrated by the activities described. It is not known whether a functional silo exists, or whether proper authority has been granted. (ITIL Service Strategy) - Service Management as a practice [ITIL - Service Management as a Practice]

7. A - A procedure describes who should carry out logically related activities, while work instructions define how activities in a procedure should be carried out at a highly detailed level. (ITIL Service Strategy) - Service Management as a practice [ITIL - Service Management as a Practice]

8. B - The Process Resources and the Process Capabilities fall under the area of Process Enablers in the Process

Model; Process Owner and Process Policy fall under the area of Process Control. (ITIL Service Design) - Service Management as a practice [ITIL - Service Management as a Practice]

9. C - A proven, effective approach which is not yet industry practice is known as a Best Practice. Once it is commonly adopted industry-wide, it becomes a state of Good Practice which is continuously improved. (General ITIL) - Service Management as a practice [ITIL - Service Management as a Practice]

10. D - In ITIL, a procedure describes logically related activities in a process, and who should carry them out. (ITIL Service Strategy) - Service Management as a practice [ITIL - Service Management as a Practice]

Knowledge Area Quiz: Service Lifecycle Practice Questions

Test Name: Knowledge Area Quiz: Service Lifecycle
Total Questions: 10
Correct Answers Needed to Pass: 7 (70.00%)
Time Allowed: 10 Minutes

Test Description

This practice test specifically targets the ITIL concepts related to the Service Lifecycle.

Test Questions

1. Which of the following ITIL Core volumes does the Service Lifecycle revolve around, and represents the policies and objectives of an organization's services?

   A. Service Strategy

   B. Service Design

   C. Service Transition

   D. Continual Service Improvement

2. Which of the statements below are not true regarding the IT Service Lifecycle?

   A. IT strategy supports the business through the design of service solutions

   B. Services are supported to maintain agreed service levels while in operation

   C. A continual improvement cycle should be adopted to ensure competitiveness

   D. IT strategy should dictate business strategy

3. Which of the following ITIL Core volumes provides guidance on achieving efficiency in the delivery and support of services to customers on a day-to-day basis?

   A. Service Design

   B. Continual Service Improvement

   C. Service Operation

   D. Service Transition

4. Taking in to account the dominant pattern of the Service Lifecycle, the

outputs of Service Design are used by what area to implement new or changed services in to production?

A. Service Transition

B. Service Design

C. Service Operation

D. Continual Service Improvement

5. Which of the following ITIL Core volumes provides guidance on managing the implementation of a new or changed service into operations?

A. Service Transition

B. Service Design

C. Continual Service Improvement

D. Service Strategy

6. Which of the following ITIL Core volumes is most useful for developing policies, guidelines, and processes across the entire Service Lifecycle?

A. Service Strategy

B. Service Design

C. Service Transition

D. Continual Service Improvement

7. Which of the following ITIL Core volumes is most heavily aligned with the Plan, Do, Check, Act (PDCA) model?

A. Continual Service Improvement

B. Service Design

C. Service Transition

D. Service Strategy

8. Which of the following ITIL Core volumes covers topics which include the development of internal and external markets, service assets and the service catalog?

A. Service Transition

B. Service Design

C. Continual Service Improvement

D. Service Strategy

9. Taking in to account the dominant pattern of the Service Lifecycle, the outputs of Service Strategy are used by

what area to plan the creation and modification of services?

A. Service Transition

B. Service Operation

C. Continual Service Improvement

D. Service Design

10. Taking in to account the dominant pattern of the Service Lifecycle, the outputs of Service Transition are used by what area to manage the day-to-day operations of a service?

A. Service Transition

B. Service Design

C. Continual Service Improvement

D. Service Operation

Knowledge Area Quiz: Service Lifecycle Answer Key and Explanations

1. A - Service Strategy represents policies and objectives, and is what the Service Lifecycle revolves around at its core. (Service Lifecycle) [ITIL - Service Lifecycle]

2. D - Business strategy should dictate IT strategy; IT strategy should not dictate business strategy. (Service Lifecycle) [ITIL - Service Lifecycle]

3. C - Service Operation provides guidance on the effectiveness and efficiency for the delivery and support of operational (day-to-day) services to customers. (Service Lifecycle) [ITIL - Service Lifecycle]

4. A - The dominant pattern of progress in the Service Lifecycle is for Service Transition to follow Service Design (Service Lifecycle) [ITIL - Service Lifecycle]

5. A - Service Transition provides guidance on managing the implementation, or transition, of a new or changed service into operations. (Service Lifecycle) [ITIL - Service Lifecycle]

6. A - Service Strategy is responsible for developing policies, guidelines, and processes across the Service Lifecycle (Service Lifecycle) [ITIL - Service Lifecycle]

7. A - Continual Service Improvement is heavily alligned with the ISO/IEC 20000 Plan, Do, Check, Act (PDCA) model. (Service Lifecycle) [ITIL - Service Lifecycle]

8. D - Service Strategy covers topics which include the development of internal and external markets, service assets and the service catalog. (Service Lifecycle) [ITIL - Service Lifecycle]

9. D - The dominant pattern of progress in the Service Lifecycle is for Service Design to follow Service Strategy (Service Lifecycle) [ITIL - Service Lifecycle]

10. D - The dominant pattern of progress in the Service Lifecycle is for Service Operation to follow Service Transition (Service Lifecycle) [ITIL - Service Lifecycle]

Knowledge Area Quiz: Generic Concepts and Definitions Practice Questions

Test Name: Knowledge Area Quiz: Generic Concepts and Definitions
Total Questions: 10
Correct Answers Needed to Pass: 7 (70.00%)
Time Allowed: 10 Minutes

Test Description

This practice test focuses on ITIL Generic Concepts and Definitions.

Test Questions

1. Under ITIL, the organization or entity responsible for the delivery of a service to a customer is known as what?

   A. Supplier

   B. Internal Market

   C. Vendor

   D. Service Provider

2. Which of the following items describes to a customer the services they will be provided, along with the expected level of service, roles and responsibilities?

   A. Operational Level Agreeement (OLA)

   B. Service Level Agreement (SLA)

   C. Service Specification

   D. Service Requirements

3. What item is a result of designing a new service, planning a major change to an existing service, or removing an existing service; and is relied on by Service Transition for its detailed requirements?

   A. Service Portfolio

   B. Service Level Agreement (SLA)

   C. Service Design Package (SDP)

   D. Quality Management Plan

4. Under Configuration Management in ITIL, this system holds information on all of an organization's Configuration Items (CIs).

A. Configuration Item (CI)

B. Configuration Management System (CMS)

C. Service Knowledge Management System (SKMS)

D. Service Catalogue

5. What term best reflects the uncertainty of an outcome, including negative threats, and positive opportunities?

A. Business Case

B. Risk

C. Procedure

D. Threat

6. A secure repository, where authorized versions of all media CIs are protected and stored, is known as what?

A. Definitive Media Library (DML)

B. Configuration Management Database (CMDB)

C. Definitive Hardware Store (DHS)

D. Configuration Management System (CMS)

7. What term in ITIL best reflects an unexpected interruption or reduction in the quality of an IT Service?

A. Problem

B. Service Request

C. Incident

D. Alert

8. What term best reflects a temporary method of resolving an issue, difficulty, or service interruption?

A. Workaround

B. Known Error

C. Incident

D. Service Request

9. What is the term used for the underpinning agreement decided on which defines the services to be provided; and is established between groups or departments which are internal to a service provider's organization?

A. SLA

B. RFC

C. Underpinning Contract

D. OLA

10. Of the items listed below, where would information relating to an organization's Service CIs be stored?

A. DML

B. CDB

C. CMS

D. KMS

Knowledge Area Quiz: Generic Concepts and Definitions Answer Key and Explanations

1. D - A Service Provider reflects the organization or entity responsible for the delivery of a service to a customer . (Service Design) - Generic Concepts and Definitions [ITIL - Generic Concepts and Definitions]

2. B - The SLA describes to the custom the services they will be provided, along with the expected level of service, roles and responsibilities. (Service Design) - Generic Concepts and Definitions [ITIL - Generic Concepts and Definitions]

3. C - The Service Design Package (SDP) is created when designing a new service, planning a major change to an existing service, or removing an existing service; it is relied on heavily by Service Transition. (Service Design) - Generic Concepts and Definitions [ITIL - Generic Concepts and Definitions]

4. B - Under ITIL, the Configuration Management System is the overarching platform holding information on all of an organization's Configuration Items (CIs). (Service Transition) - Generic Concepts and Definitions [ITIL - Generic Concepts and Definitions]

5. B - A Risk represents an uncertain outcome, and can be positive or negative (opportunity or threat). (Service Lifecycle) - Generic Concepts and Definitions [ITIL - Generic Concepts and Definitions]

6. A - The Definitive Media Library (DML) stores authorized versions of all media CIs, including master copies of all controlled software in an organization. (Service Transition) - Generic Concepts and Definitions [ITIL - Generic Concepts and Definitions]

7. C - An incident is an unexpected interruption or reduction in the quality of an IT Service (Service Operation) - Generic Concepts and Definitions [ITIL - Generic Concepts and Definitions]

8. A - A Workaround provides a temporary means of resolving an issue for which an underlying root cause has not yet been resolved. (Service Operation) - Generic Concepts and Definitions [ITIL - Generic Concepts and Definitions]

9. D - An Operational Level Agreement (OLA) is the term used to describe a contract established between groups

or departments which are internal to a service provider's organization. (Service Design) - Generic Concepts and Definitions [ITIL - Generic Concepts and Definitions]

10. C - The Configuration Management System (CMS) would store information on an organization's Service Configuration Items (CIs). (Service Transition) - Generic Concepts and Definitions [ITIL - Generic Concepts and Definitions]

ITIL Foundation
Practice Exam 2
Practice Questions

Test Name: ITIL Foundation Practice Exam 2
Total Questions: 40
Correct Answers Needed to Pass: 30 (75.00%)
Time Allowed: 60 Minutes

Test Description

This is the second cumulative ITIL Foundation test which can be used as an indicator for overall performance. This practice test includes questions from all ITIL categories.

Test Questions

1. Under ITIL, how is the value of a service defined?

   A. Through revenue and profit generated

   B. Through business outcomes and customer perception

   C. Through consistency and quality

   D. Through service cost and demand

2. What generic Service Management term is used to describe technology which allows users to find resolution to support requests without the assistance of Service Desk personnel, often relying on web based access to accommodate Service Requests?

   A. Self-Help technology

   B. Incident Management System

   C. Known Error Database

   D. Virtual Service Desk

3. From the customer's perspective, what are the two most important components which make up the value of a service?

   A. Utility and Warranty

   B. Resources and Capabilities

   C. Utility and Resources

   D. Design and Capacity

4. In ITIL, what is the generic sequence of events which take place in a Process Model?

   I. Data enters

II. Data is processed
III. Data is measured and reviewed
IV. Data is output

A. I, III, II, IV

B. I, II, IV, III

C. I, II, III, IV

D. I, IV, II, III

5. What type of Service Desk reduces the number of service desk staff by merging all of them into a single location, is generally more efficient and cost effective, and typically contain a high level of skill?

A. Local Service Desk

B. Centralized Service Desk

C. Virtual Service Desk

D. Follow the Sun

6. Which of the following ITIL Core volumes is most useful for developing policies, guidelines, and processes across the entire Service Lifecycle; and places a strong focus on Financial Management and Service Portfolio Management?

A. Service Strategy

B. Service Design

C. Service Transition

D. Continual Service Improvement

7. What aspect of Service Design is concerned with the management and control of services throughout their lifecycle via service management systems and related tools?

A. Technology architectural design

B. Process Design

C. Measurement Design

D. Service Portfolio Management

8. Under ITIL, an external third party who is necessary to support the components involved in providing a service is known as what?

A. Supplier

B. Internal Market

C. Service Provider

D. Customer

9. What type of technology allows authorized support groups to take control of a user's desktop from a different physical location?

A. Event management tools

B. Remote control tools

C. Discovery, Deployment, and Licensing Technologies

D. Diagnostic utilities

10. SLAs, OLAs, and Underpinning Contracts are defined within Service Design by which ITIL process?

A. Service Level Management

B. Change Management

C. Service Strategy

D. Continual Service Improvement

11. What term best reflects a Service Management product's ability to adequately perform?

A. Capacity

B. Scalability

C. Continuity

D. Security

12. In the RACI model, what role represents the people who are kept updated on the progress of activities?

A. Responsible

B. Accountable

C. Consulted

D. Informed

13. What activity is an organization performing when it plans and regulates a process with the goal of attaining effective and consistent results?

A. Process improvement

B. Process execution

C. Process analysis

D. Process control

14. Which of the statements below are true regarding service assets?

I. The performance potential of customer assets increases as service potential is increased
II. Increased customer performance potential results in less demand for the scale or scope of a service

A. I and II

B. I

C. II

D. Both statements are false

15. For which ITIL Core volume is the Service Desk function most closely aligned?

A. Service Strategy

B. Service Design

C. Service Operation

D. Service Transition

16. Which Service Design process ensures that information regarding services which run in the live environment is current and accurate?

A. Service Catalogue Management

B. Service Level Management

C. Availability Management

D. Service Improvement

17. To ensure consistency, professionalism, and efficiency when in contact with the customer, the Service Desk may be provided with set procedures based on questionnaires and standard responses. These standardized procedures, questionnaires and responses used by the Service Desk are collectively known as:

A. Screenplays

B. Scripts

C. Drafts

D. Speeches

18. What term best reflects where an organization would store all services, including services being developed, services currently offered, and services which have been retired?

A. Service Catalogue

B. Service Pipeline

C. Service Specification

D. Service Portfolio

19. What role is responsible for negotiating levels of service with the customer through SLAs and SLRs and ensures Underpinning Contracts are in agreement with existing SLA and SLR targets?

A. IT Financial Manager

B. Service Level Manager

C. Demand Manager

D. Service Catalogue Manager

20. The measure of how quickly a service can be restored to normal operating status following a failure or interruption is known as what?

A. Availability

B. Reliability

C. Maintainability

D. Serviceability

21. What statement below best describes the concept of a "Process" in ITIL?

A. A logical concept referring to people and automated measures that execute an activity.

B. A means of delivering value to customers by facilitating outcomes they want to achieve, without the ownership of specific costs and risks.

C. The activity of planning and regulating a set of activities, with the objective of consistent outcomes.

D. A set of coordinated activities combining and implementing resources and capabilities in order to produce an outcome which creates value for the customer or stakeholder.

22. In the Service Lifecycle, what is the dominant pattern of progress between the process areas?

I. Service Design
II. Service Transition
III. Service Operation
IV. Service Strategy

A. I, IV, II, III

B. IV, I, III, II

C. IV, I, II, III

D. I, II, IV, III

23. What Service Transition process must maintain updated and accurate configuration records, in addition to defining and controlling the components which make up an organization's services and infrastructure?

A. Availability Management

B. Service Asset and Configuration Management

C. Change Management

D. Service Measurement

24. Which of the following items describes Service Desk personnel with low incident handling and high resolution rate?

A. Follow the Sun

B. Technical Skill Level

C. Local Service Desk

D. Basic Skill Level

25. What role assists with Business Impact Analyses by defining the level of control and protection for an organization's information assets; and performs tests which seek to identify information vulnerabilities?

A. Availability Manager

B. Capacity Manager

C. Security Manager

D. Service Continuity Manager

26. Of the items listed below, which serves as the blueprint for service management functions and processes, and specify how service assets interact with customer assets to create value?

A. Service Portfolio

B. Procedure

C. Service Model

D. Service Catalogue

27. An addition, modification, or removal of a supported service component to resolve an error found in the service, is known as what?

A. Configuration Item (CI)

B. Service Change

C. Change Request

D. Work Instruction

28. Under CSI, an established starting data point used to for comparison in the future is known as what?

A. Metric

B. Event

C. Baseline

D. Validation

29. Which of the statements below are true regarding the Service Lifecycle?

I. The Service Lifecycle should provide the appropriate structure and stability to service management capabilities through sound principles and tools.
II. The Service Lifecycle should provide the basis for measurement, learning, and improvement.

A. I and II

B. I

C. II

D. None of the above are true

30. Under ITIL, a process takes one or more _______ and turns them in to defined _______.

A. stakeholders, customers

B. functions, roles

C. inputs, outputs

D. service assets, customer assets

31. What role is responsible for ensuring coordination between the build, test, and release teams; plans service roll-outs; and manages the installation of new or upgraded hardware?

A. Configuration Manager

B. Service Asset Manager

C. Change Manager

D. Release and Deployment Manager

32. Which of the following ITIL Core volumes represent the progressive phases of the Service Lifecycle by implementing strategy and representing change?

I. Service Design
II. Continual Service Improvement
III. Service Transition
IV. Service Operation

A. I, II, III

B. I, III, IV

C. II, III, IV

D. II, III

33. Under Continual Service Improvement (CSI), what type of metrics are typically captured in the form of Key Performance Indicators (KPIs), Critical Success Factors (CSFs) and relate to processes within Service Management?

A. Technology Metrics

B. Process Metrics

C. Baseline Metrics

D. Service Metrics

34. Which item below best reflects a self-contained, specialized organizational unit which has their own body of knowledge, and optimizes their work methods by focusing on outcome?

A. Function

B. Role

C. Best Practice

D. Service

35. Which of the statements below can be considered true / correct?

I. An Error Report is an incident which reflects faults or complaints about the service
II. A Service Request is an incident which involves a failure in the IT infrastructure

A. I

B. II

C. I and II

D. None of the above

36. Which of the following processes are not owned by the ITIL Core volume: Continual Service Improvement (CSI)?

A. Service Measurement

B. Service Reporting

C. Service Improvement

D. Service Level Management

37. What term in ITIL best reflects the underlying root cause for one or more incidents?

A. Service Request

B. Event

C. Alert

D. Problem

38. What model, utilized under Continual Service Improvement (CSI), consists of 6 steps which include embracing the vision, assessing the current situation, agreeing on priorities, planning to achieve quality, verifying metrics, and ensuring momentum for improvement is maintained?

A. Deming Cycle

B. CSI Model

C. PDCA Model

D. Service Lifecycle

39. What role is responsible for the sizing and performance testing of new services and systems?

A. Capacity Manager

B. Configuration Manager

C. Availability Manager

D. Service Continuity Manager

40. The DML, which stores and protects the definitive versions of all media CIs, stands for what?

A. Definitive Media Library

B. Definitive Microcomputer Lab

C. Detailed Media Lab

D. Detailed Meta Library

ITIL Foundation Practice Exam 2 Answer Key and Explanations

1. B - Value is defined by the business outcomes and customer perception. (Service Strategy) - Key Principles and Models [ITIL - Key Principles and Models]

2. A - This approach describes the concept of Self-Help technologies. (Service Management) - Technology and Architecture [ITIL - Technology and Architecture]

3. A - From the customer's perspective, the value of a service is represented by fitness for purpose (Utility) and fitness for use (Warranty). (Service Management) - Generic Concepts and Definitions [ITIL - Generic Concepts and Definitions]

4. B - The generic sequence of events in the ITIL Process Model is for data to enter, be processed, output, and finally measured. (ITIL Service Design) - Service Management as a practice [ITIL - Service Management as a Practice]

5. B - The characteristics above describe a Centralized Service Desk. (Service Operation) - Selected Functions [ITIL - Selected Functions]

6. A - Service Strategy is responsible for developing policies, guidelines, and processes across the Service Lifecycle (Service Lifecycle) [ITIL - Service Lifecycle]

7. D - In Service Design, Service Portfolio Management is concerned with the management and control of services throughout their lifecycle. (Service Design) - Key Principles and Models [ITIL - Key Principles and Models]

8. A - A Supplier is an external third party who provides support to the components involved in delivering a service . (Service Design) - Generic Concepts and Definitions [ITIL - Generic Concepts and Definitions]

9. B - Remote control tools allow authorized support groups to take control of a user's desktop from a different physical location and include software such as Microsoft Remote Desktop. (Service Management) - Technology and Architecture [ITIL - Technology and Architecture]

10. A - CSI and Service Strategy are ITIL Core volumes, not processes. Service Level Management is the process which covers SLAs, OLAs, and

Underpinning Contracts. (Service Design) - Selected Processes [ITIL - Selected Processes]

11. A - This concept is best represented by the term: Capacity. (Service Management) - Technology and Architecture [ITIL - Technology and Architecture]

12. D - The "Informed" role of the RACI model are those people who are kept updated on the progress of activities. (Service Management / Service Lifecycle) - Selected Roles [ITIL - Selected Roles]

13. D - Process control is concerned with the planning and regulating of processes with the goal of producting effective, consistent results. (Service Design) - Key Principles and Models [ITIL - Key Principles and Models]

14. B - Increased customer performance potential results in more demand, not less, for the scale or scope of a service.(ITIL Service Strategy) - Service Management as a practice [ITIL - Service Management as a Practice]

15. C - Although the Service Desk may play a role in supporting a number of processes and ITIL areas, it is most closely aligned with Service Operation. (Service Operation) - Selected Functions [ITIL - Selected Functions]

16. A - Service Catalogue Management ensures that information regarding services which run in the live environment is current and accurate. Service Improvement is a process covered by Continual Service Management. (Service Design) - Selected Processes [ITIL - Selected Processes]

17. B - Scripts can be provided to the Service Desk in order to provide standardized procedures based on questionnaires and responses, and add to the efficiency and professionalism of the Service Desk. (Service Operation) - Selected Functions [ITIL - Selected Functions]

18. D - The Service Portfolio consists of services in development, services currently being offered, and services which have been retired. The Service Pipeline consists entirely of services in development. The Service Catalogue consists of services currently being offered. (Service Lifecycle) - Generic Concepts and Definitions [ITIL - Generic Concepts and Definitions]

19. B - The Service Level Manager performs the activities above, and also identifies the key stakeholders impacted by service levels. (Service

Design) - Selected Roles [ITIL - Selected Roles]

20. C - Under Availability Management, the measure of how quickly a service can be restored to normal operating status following a failure or interruption is known as Maintainability. (Service Design) - Selected Processes [ITIL - Selected Processes]

21. D - ITIL describes a Process as "A set of coordinated activities combining and implementing resources and capabilities in order to produce an outcome which creates value for the customer or stakeholder.". (ITIL Service Strategy) - Service Management as a practice [ITIL - Service Management as a Practice]

22. C - The general sequence of the Service Lifecycle is Service Strategy, Service Design, Service Transition, Service Operation. (Service Lifecycle) [ITIL - Service Lifecycle]

23. B - are Service Transition processes. Of these, Service Asset and Configuration Management is the correct process. (Service Transition) - Selected Processes [ITIL - Selected Processes]

24. B - Low incident handling and high resolution rate is categorized by a Technical Skill Level. (Service Operation) - Selected Functions [ITIL - Selected Functions]

25. C - The Security Manager performs the activities above, and also develops and maintains a Security Policy and security plans. (Service Design) - Selected Roles [ITIL - Selected Roles]

26. C - A Service Model specifies how service assets interact with customer assets and provides a blue print for the service management processes. (Service Lifecycle) - Generic Concepts and Definitions [ITIL - Generic Concepts and Definitions]

27. B - A Service Change best describes the addition, modification, or removal of a service component, which may take place for reasons including corrective actions to resolve errors. (Service Transition) - Generic Concepts and Definitions [ITIL - Generic Concepts and Definitions]

28. C - A baseline is an initial starting point used to make comparisons against in the future. (Continual Service Improvement) - Key Principles and Models [ITIL - Key Principles and Models]

29. A - Both of the statements are true regarding the Service Lifecycle.

(Service Lifecycle) [ITIL - Service Lifecycle]

30. C - Under ITIL, a process takes one or more inputs and turns them in to defined outputs. (ITIL Service Strategy) - Service Management as a practice [ITIL - Service Management as a Practice]

31. D - The Release and Deployment Manager is responsible for the activities above, and manages all aspects of the release process within an organization. (Service Transition) - Selected Roles [ITIL - Selected Roles]

32. B - The ITIL Core volumes which represent change and transformation are Service Design, Service Transition, and Service Operation. Continual Service Improvement accomodates learning and improvement, prioritizing improvement programs and projects. (Service Lifecycle) [ITIL - Service Lifecycle]

33. B - Process Metrics are captured in the form of Key Performance Indicators (KPIs), Critical Success Factors (CSFs) and relate to processes within Service Management. (Continual Service Improvement) - Key Principles and Models [ITIL - Key Principles and Models]

34. A - Under ITIL, this statement best describes a Function. (ITIL Service Strategy) - Service Management as a practice [ITIL - Service Management as a Practice]

35. A - A Service Request is an incident which does not involve a failure in the IT infrastructure. The first statement is correct. (Service Operation) - Selected Functions [ITIL - Selected Functions]

36. D - Service Level Management is not owned by CSI, is a key process of Service Design. (Continual Service Improvement) - Selected Processes [ITIL - Selected Processes]

37. D - In ITIL, a Problem is the undertlying root cause for one or more incidents. (Service Operation) - Generic Concepts and Definitions [ITIL - Generic Concepts and Definitions]

38. B - The 6 steps describe the components of the CSI Model, which is an iterative improvement process under Continual Service Improvement (CSI). (Continual Service Improvement) - Key Principles and Models [ITIL - Key Principles and Models]

39. A - The Capacity Manager is responsible for sizing and performance testing of new services and systems. The term "sizing" alludes to capacity. (Service Design) - Selected Roles [ITIL - Selected Roles]

40. A - The DML stands for "Definitive Media Library" which stores all media CIs. (Service Transition) - Generic Concepts and Definitions [ITIL - Generic Concepts and Definitions]

Knowledge Area Quiz:
Key Principles and Models
Practice Questions

Test Name: Knowledge Area Quiz: Key Principles and Models
Total Questions: 10
Correct Answers Needed to Pass:
7 (70.00%)
Time Allowed: 10 Minutes

Test Description

This practice test targets the Key Principles and Models of ITIL.

Test Questions

1. Service Portfolio Management, identification of business requirements, technology architectural design, process design, and measurement design are five critical aspects of what stage in the ITIL Service Lifecycle?

    A. Service Strategy

    B. Service Design

    C. Service Transition

    D. Service Operation

2. An integrated and holistic approach to Service Design should incorporate input from the primary elements of an IT service. Which of the following is not one of the five major aspects of service design?

    A. Management information systems and tools

    B. Technology and management architecture

    C. Cost and timeline

    D. Processes

3. The Continual Service Improvement (CSI) Approach summarizes the ongoing improvement cycle. This approach starts with what important question?

    A. Where are we going?

    B. What is the vision?

    C. What services are we offering?

    D. Are we under or over budget?

4. What Service Operation principle focuses on the development and refinement of standard IT

management processes which lead to services that are available and perform consistently?

A. Responsiveness

B. Process Management

C. Cost of Service

D. Stability

5. Under Continual Service Improvement, what are the four stages of the Deming Cycle, relied on for steady, ongoing improvement?

A. Plan, Design, Build, Deploy

B. Design, Build, Deploy, Support

C. Design, Check, Act, Fix

D. Plan, Do, Check, Act

6. Under Continual Service Improvement, all four stages of the Deming Cycle are applied during "implementation". What stages of the Deming Cycle are applied during "ongoing improvement" to monitor, measure, and implement initiatives?

I. Plan
II. Do
III. Check
IV. Act

A. I and II

B. III and IV

C. I and III

D. II and IV

7. What stage of the Deming Cycle requires a comparison of the service improvements which have been implemented, against the metrics of success?

A. Plan

B. Do

C. Check

D. Act

8. Under Continual Service Improvement (CSI), what are the four basic objectives of Service Measurement?

A. Validate, Direct, Justify, Intervene

B. Validate, Measure, Analyze, Deploy

C. People, Process, Products, Partners

D. Plan, Do, Check, Act

9. Applied heavily under Continual Service Improvement (CSI), what does the PDCA Model stand for?

A. Plan, Do, Check, Act

B. Plan, Design, Confirm, Act

C. Produce, Deploy, Change, Accept

D. Produce, Design, Check, Accept

10. Taking in to account the use of baseline data under Continual Service Improvement (CSI), which of the statements below are true?

I. If the integrity of measurement data is questionable, it is better to not have any data at all.
II. If a baseline has not yet been established, the first measurements will immediately become the baseline

A. I and II

B. I

C. II

D. None of the above are true.

Knowledge Area Quiz: Key Principles and Models Answer Key and Explanations

1. B - These activities are critical to Service Design. On its own, Service Portfolio Management is a key process associated with Service Strategy; when applied in conjunction with the four additional activities stated, it becomes a critical aspect of Service Design. (Service Design - Key Principles and Models) [ITIL - Key Principles and Models]

2. C - Cost and timeline are not one of the major aspects of service design. These five aspects are: management information systems and tools, technology and management architecture, processes, measurement methods and metrics, and service solutions for new or changed services. [ITIL - Key Principles and Models]

3. B - The CSI Approach uses a series of questions to gather and analyze information about existing IT services and subsequently make plans to develop and improve services. Touching all areas of the Service Lifecycle, these questions are: What is the vision? Where are we now? Where do we want to be? How do we get there? Did we get there? [ITIL - Key Principles and Models]

4. D - "Stability" focuses on the development and refinement of standard IT management processes which leading to services that are available and perform consistently. (Service Operation) - Key Principles and Models [ITIL - Key Principles and Models]

5. D - Plan, Do, Check, Act are the four stages of the Deming Cycle which are heavily relied on by Continual Service Improvement. (Continual Service Improvement) - Key Principles and Models [ITIL - Key Principles and Models]

6. B - When practicing ongoing improvement, CSI relies on the Check and Act stages of the Deming Cycle to monitor, measure, review and implement initiatives. (Continual Service Improvement) - Key Principles and Models [ITIL - Key Principles and Models]

7. C - The "Check" stage requires a comparison of the service improvements which have been implemented, against the metrics of success. (Continual Service Improvement) - Key Principles and Models [ITIL - Key Principles and Models]

8. A - The four basic objectives of Service Measurement under CSI are Validate, Direct, Justify, Intervene. (Continual Service Improvement) - Key Principles and Models [ITIL - Key Principles and Models]

9. A - The PDCA Model, otherwise known as the Deming Cycle, stands for Plan, Do, Check, Act. (Continual Service Improvement) - Key Principles and Models [ITIL - Key Principles and Models]

10. C - Even if the integrity of measurement data is questionable, it is better than not having any data at all. At the very least, there will be data to question. (Continual Service Improvement) - Key Principles and Models [ITIL - Key Principles and Models]

Knowledge Area Quiz: Selected Processes Practice Questions

Test Name: Knowledge Area Quiz: Selected Processes
Total Questions: 10
Correct Answers Needed to Pass: 7 (70.00%)
Time Allowed: 10 Minutes

Test Description

This practice test focuses on selected ITIL processes.

Test Questions

1. Cherryvale Logistics offers a number of services to its users. The finance department has recently asked for an expansion of the core business hours its database is available. Which process is responsible for negotiating this change in the level of service?

   A. Operations management

   B. IT Service Management

   C. Business relationship management

   D. Service level management

2. Each department at regional construction firm is charged $0.01 per page printed. These charges are deducted from each department's shared services budget. What process is responsible for developing and implementing this charging system?

   A. Financial management for IT services

   B. Business service management

   C. Accounting controls

   D. Cost center management

3. A new operating system platform has been implemented at Lincoln Utilities during the weekend change window. However, it has been discovered that the new platform is affecting the availability of the organization's billing system. Which of the following will be implemented to address this?

   A. Remediation plan

   B. Patch management

   C. Change control

   D. Rollback plan

4. What Service Design process manages the performance and capacity of services and resources, and maintains a Capacity Plan which reflects the current and future needs of the organization?

   A. Service Level Management

   B. Capacity Management

   C. Availability Management

   D. Service Catalogue Management

5. Big-bang and phased approaches, push and pull approaches, and automated versus manual approaches are all concepts which fall under what Service Transition process?

   A. Service Asset and Configuration Management

   B. Change Management

   C. Release and Deployment Management

   D. Availability Management

6. Which of the following processes is not owned by the ITIL Core volume "Service Operation"?

   A. Incident Management

   B. Problem Management

   C. Change Management

   D. Event Management

7. What Service Operation process focuses on restoring the normal operation of a service as quickly as possible while minimizing any impact to the operation of the business?

   A. Incident Management

   B. Problem Management

   C. Change Management

   D. Event Management

8. What Service Operation process focuses on the type and extent of data accessible to a user, identify of a user, and the rights of a user?

   A. Access Management

   B. Incident Management

   C. Event Management

   D. Availability Management

9. What Service Design process completes regular Business Impact Analyses (BIAs) and ensures services remain available following disasters?

A. Capacity Management

B. Availability Management

C. Service Continuity Management

D. Service Level Management

10. Under Availability Management, what term best describes the measure of how long a service, component, or CI can perform its specific, agreed function without interruption?

A. Availability

B. Maintainability

C. Serviceability

D. Reliability

Knowledge Area Quiz: Selected Processes Answer Key and Explanations

1. D - Service level management is the process responsible for developing, negotiating, monitoring, reporting, and reviewing IT service targets, such as the specific hours a service is guaranteed to be available. [ITIL - Selected Processes]

2. A - Financial management for IT services ensures that the appropriate level of funding to design ,develop, and deliver IT services is secured. Funding models can include charging, requiring customers to pay for the services that they use. [ITIL - Selected Processes]

3. A - The remediation plan will be implemented to address the service availability with the operating system change. Remediation plans are developed prior to implementation of a change, and may include options, processes, and trigger points to indicate how a malfunctioning change will be dealt with. [ITIL - Selected Processes]

4. B - As the name suggests, Capacity Management manages the performance and capacity of services and resources, and maintains a Capacity Plan which reflects the current and future needs of the organization. (Service Design) - Selected Processes [ITIL - Selected Processes]

5. C - The concepts describe approaches towards the release and deployment of service components, and fall under Release and Deployment Management. (Service Transition) - Selected Processes [ITIL - Selected Processes]

6. C - Change Management is owned by the ITIL Core volume "Service Transition". (Service Operation) - Selected Processes [ITIL - Selected Processes]

7. A - Incident Management focuses on restoring the normal operation of a service as quickly as possible while minimizing any impact to the operation of the business. (Service Operation) - Selected Processes [ITIL - Selected Processes]

8. A - Access Management focuses on the type and extent of data accessibile to a user, identify of a user, and the rights of a user. Availability Management is not owned by Service Operation. (Service Operation) - Selected Processes [ITIL - Selected Processes]

9. C - Business Impact Analyses (BIAs) ensures services remain available following disasters. (Service Design) - Selected Processes [ITIL - Selected Processes]

10. D - Under Availability Management, Reliability describes the measure of how long a service, component, or CI can perform its specific, agreed function without interruption. (Service Design) - Selected Processes [ITIL - Selected Processes]

Knowledge Area Quiz: Selected Functions Practice Questions

Test Name: Knowledge Area Quiz: Selected Functions
Total Questions: 10
Correct Answers Needed to Pass: 7 (70.00%)
Time Allowed: 10 Minutes

Test Description

This practice test focuses on selected ITIL functions.

Test Questions

1. What type of Service Desk is co-located in or around the user communities, is clearly visible, and may result in inefficiencies and higher costs if the staff waits to deal with incidents?

   A. Local Service Desk

   B. Centralized Service Desk

   C. Virtual Service Desk

   D. Follow the Sun

2. What type of Service Desk has personnel spread across a number of geographical locations, relies heavily on technology and corporate support tools, and may need safeguards to ensure consistent service quality?

   A. Local Service Desk

   B. Centralized Service Desk

   C. Virtual Service Desk

   D. Follow the Sun

3. What ITIL function plays a dual role as the custodian of technical knowledge and expertise, and provides actual resources to support the Service Management Lifecycle?

   A. Service Desk

   B. Technical Management

   C. Application Management

   D. IT Operations Management

4. What ITIL function manages an organization's applications throughout their lifecycle?

   A. Service Desk

B. Technical Management

C. Application Management

D. IT Operations Management

5. How does the Service Desk reduce the workload on other IT departments within an organization?

A. By acting as the initial point of contact

B. By intercepting user questions which are easily answered before they reach specialist personnel

C. By escalating support calls to second and third-line support only when needed

D. All of these responses / All of the above

6. Which of the following items describes Service Desk personnel with high incident handling and low resolution rate?

A. Follow the Sun

B. Technical Skill Level

C. Virtual Service Desk

D. Basic Skill Level

7. Of the items listed below, which is least likely to be an example of a Key Performance Indicator for the Service Desk?

A. What was the First-line support resolution rate?

B. What was the number of incidents resulting from changes?

C. What was the average time to resolve an incident?

D. What percentage of support calls were functionally escalated within 6 minutes of not being resolved?

8. Which of the statements below can be considered true / correct?

I. The Service Desk should act as the main source of information for users, and should provide users with information regarding current and expected errors.
II. The Service Desk should be capable of providing users with information regarding SLA provisions, new and

existing services, and order procedures.

A. I

B. II

C. I and II

D. None of the above

9. What types of incidents should be logged by the Service Desk?

A. Only incidents which cannot be resolved by the Service Desk personnel

B. Only incidents which require functional escalation

C. All incidents excluding Service Requests

D. All incidents should be logged by the Service Desk

10. In regards to the Service Desk, a measurement of "What percentage of calls were answered within 45 seconds?" is an example of a:

A. Critical Success Factor

B. Key Performance Indicator

C. Best Practice

D. High Priority

Knowledge Area Quiz: Selected Functions Answer Key and Explanations

1. A - The characteristics above describe a Local Service Desk. (Service Operation) - Selected Functions [ITIL - Selected Functions]

2. C - The characteristics above describe a Virtual Service Desk. (Service Operation) - Selected Functions [ITIL - Selected Functions]

3. B - The Technical Management function plays a dual role as the custodian of technical knowledge and expertise, and provides actual resources to support the Service Management Lifecycle. (Service Operation) - Selected Functions [ITIL - Selected Functions]

4. C - The Application Management function manages an organization's applications throughout their lifecycle and will contain staff who are logically a part of IT Operations Management. (Service Operation) - Selected Functions [ITIL - Selected Functions]

5. D - All of the statements listed are reasons why the Service Desk reduces the workload on other IT departments within the organization. (Service Operation) - Selected Functions [ITIL - Selected Functions]

6. D - High incident and low resolution is categorized by a Basic Skill Level. (Service Operation) - Selected Functions [ITIL - Selected Functions]

7. B - Of the options provided, the number of incidents resulting from changes would more than likely reflect a KPI from the Change Management process. (Service Operation) - Selected Functions [ITIL - Selected Functions]

8. C - Both of the statements are true and correct in regards to the Service Desk's duty to provide information to users. (Service Operation) - Selected Functions [ITIL - Selected Functions]

9. D - As the correct response indicates, all incidents should be logged by the Service Desk. (Service Operation) - Selected Functions [ITIL - Selected Functions]

10. B - This measurement is an example of a Key Performance Indicator which would apply to the Service Desk. (Service Operation) - Selected Functions [ITIL - Selected Functions]

ITIL Foundation
Practice Exam 3
Practice Questions

Test Name: ITIL Foundation Practice Exam 3
Total Questions: 40
Correct Answers Needed to Pass: 30 (75.00%)
Time Allowed: 60 Minutes

Test Description

This is the third cumulative ITIL Foundation test which can be used as an indicator for overall performance. This practice test includes questions from all ITIL areas.

Test Description:

1. Markham Management has contracted with Diamond Data Systems to provide 24x7x365, 4-hour response, on-site service for Markham's networking equipment. This contract is renewed annually, and Markham may add equipment to the contract at any time. What role does Diamond Data Systems play in this scenario?

   A. Supplier

   B. Underpinning Contract

   C. Service Provider

   D. Warranty Provider

2. Under ITIL, what statement below describing a customer is false?

   A. A customer pays for the service on behalf of the user

   B. A customer cannot be the business itself

   C. A customer can also be a user

   D. A customer can be anyone who makes use of an IT service to achieve their specific outcomes

3. In what ITIL Core volume is Service Level Management most widely associated with and practiced?

   A. Service Strategy

   B. Service Design

   C. Service Transition

   D. Continual Service Improvement

4. A large storage array has been installed to support a new digital locker service offered by coblus.com. Which of the following terms best describes this array?

A. A. Hardware

B. B. Resource

C. C. SAN

D. D. NAS

5. Which of the statements below is not true about the value proposition of services?

A. Service assets provide a source of value, while customer assets act as the recipient

B. Customer assets provide a source of value, while service assets acts as the recipient

C. Improving the design of services increases customer performance

D. Improving the design of services reduces the risk of variations of customer assets

6. Which of the following ITIL Core volumes covers the design principles and methods for converting strategic objectives in to portfolios of services and service assets?

A. Continual Service Improvement

B. Service Transition

C. Service Design

D. Service Strategy

7. Which of the following is not a key metric associated with Service Level Management?

A. SLA Targets Missed

B. SLA Breaches in Underpinning Contracts

C. Incidents due to Capacity Shortages

D. Customer Satisfaction of SLA Achievements

8. Which of the following is an activity of the Service Desk?

I. To function as the first point of contact for the customer
II. To investigate the underlying cause of service interruptions for the customer

III. To determine the root cause of incidents as they occur

A. Item I

B. Item II

C. Item III

D. All of these responses / All of the above

9. Which of the following ITIL Core volumes provides guidance on managing the complexity associated with changes to service; and relies most heavily on practices derived from release management, program management, and risk management?

A. Service Design

B. Continual Service Improvement

C. Service Strategy

D. Service Transition

10. What statement below best describes the concept of a "Function" in ITIL?

A. A set of specialized organizational capabilities for providing value to customers in the form of services.

B. A logical concept referring to people and automated measures that execute a defined process, an activity, or a combination thereof.

C. A means of delivering value to customers by facilitating outcomes they want to achieve, without the ownership of specific costs and risks.

D. A team, unit, or person that performs tasks related to a specific process.

11. An engineering team is considered what type of asset?

A. Capability

B. Staff

C. Resource

D. Human Resources

12. What term best reflects a Service Management product's ability to maintain data integrity?

A. Capacity

B. Scalability

C. Continuity

D. Security

13. Which of the following is not a Core volume of the ITIL Library?

A. Service Strategy

B. Service Transition

C. Service Lifecycle

D. Service Operation

14. Of the items below, what information would the Service Desk provide to the IT management of an organization?

A. The number of calls handled by the Service Desk overall, and by workstation

B. The number of resolved problems, and the reduction in related incidents

C. The number of successfully implemented changes

D. The cost of implemented changes

15. What item is responsible for storing only those services which are in development?

A. Service Catalogue

B. Service Pipeline

C. Service Specification

D. Service Portfolio

16. Under Service Design, what process is comprised of the following four key aspects: Availability, Reliability, Maintainability, and Serviceability?

A. Service Continuity Management

B. Capacity Management

C. Availability Management

D. Supplier Management

17. What type of Service Catalogue represents the customer's view and includes relationships to the business units that rely on the IT services?

A. Service Pipeline

B. Technical Service Catalogue

C. Business Service Catalogue

D. Business Case

18. The complete list of all current customer facing IT services and supporting IT services or available for deployment is the:

A. Service Catalog

B. Service Portfolio

C. Service Database

D. Service Pipeline

19. Which of the following is not a primary characteristic of processes under ITIL?

A. Processes are measurable

B. Processes have specific results

C. Processes are strategic

D. Processes respond to a specific event

20. Which of the following is not a Core volume of the ITIL Library?

A. Service Strategy

B. Service Management

C. Service Transition

D. Continual Service Improvement

21. What role is responsible for ensuring IT recovery plans are up to date and would be a key resource in the event of a fail-over to a secondary location following a disaster scenario?

A. Availability Manager

B. Capacity Manager

C. Security Manager

D. Service Continuity Manager

22. Which of the following statements regarding Service Automation are true?

I. Service Automation is limited to serving during a business's normal operating hours

II. Service Automation enables the capturing of knowledge related to a service process; beneficial for when employees are no longer a part of the organization

A. I and II

B. I

C. II

D. Both statements are false

23. What process relies heavily on the use of a CAB, which may itself include the Problem Manager, Service Level Manager, and staff from an organization's customer relations department?

A. Service Asset and Configuration Management

B. Supplier Management

C. Service Level Management

D. Change Management

24. Which of the following tasks will a good Service Desk provide?

I. Acting as the main source of information to users
II. Identifying the root cause of incidents and creating a workaround
III. To escalate incidents if they cannot be resolved within a specific amount of time
IV. Informing users about current or expected errors

A. I, II, III

B. II, III, IV

C. I, III, IV

D. All of these responses / All of the above

25. What role is responsible for receiving, logging, and prioritizing RFCs?

A. Product Manager

B. Change Authority

C. Change Manager

D. Change Advisory Board

26. __________ enables understanding and assists in articulating distinct process features. It will be applied generically as the foundation for any process.

A. The Process Model

B. Process Control

C. Work Instruction

D. Procedure

27. Under Continual Service Improvement (CSI), which of the following are objectives of Service Measurement?

I. To validate decisions made earlier
II. To justify, with factual evidence, that a specific course of action is required
III. To determine where to intervene in case changes or corrective actions are required

A. I

B. II

C. III

D. All of these responses / All of the above

28. Under Continual Service Improvement (CSI), what types of metrics are typically associated with system components, application performance, and availability?

A. Technology Metrics

B. Process Metrics

C. Baseline Metrics

D. Service Metrics

29. A portion of the IT infrastructure which is normally deployed together, and is specified in the organization's release policy, is known as what?

A. Release Unit

B. Release Group

C. Work Instruction

D. Full Release

30. What is the term for user requests which are classified as incidents, but do not involve a failure in the IT infrastructure?

A. Service Requests

B. Non-standard Changes

C. Error Reports

D. Priority Requests

31. Which of the following would not be a likely key metric for the Incident Management process?

A. Total number of incidents logged in the past month

B. Total number of major incidents

C. Average cost to resolve each incident

D. Time to resolve underlying root cause and prevent recurring incidents

32. What role may be delegated to Service Desk and / or IT Service Management resources, and ensures that auto responses from relevant services are properly defined and executed?

A. Incident Manager

B. Problem Manager

C. Alert Manager

D. Event Manager

33. Which of the statements below are true?

I. An Event is a notification caused by a deviation in the performance of the infrastructure and is created by a user.

II. An Alert is a warning or notice about a change or failure that has occurred, and is controlled by System Management tools and the Event Management Process.

A. I and II

B. I

C. II

D. None of the above area true

34. What Service Strategy process optimizes the use of IT resources, reduces excess capacity, balances the supply and demand of resources?

A. Demand Management

B. Financial Management

C. Capacity Management

D. Service Portfolio Management

35. The Service Desk is flooded with calls from customers who can no longer work due to a system failure. By further questioning callers, it is apparent that a system-wide update occurred which crashed the central server. For which of the following activities is the Service Desk not responsible for?

A. Investigating and identifying the cause of the failure

B. Categorizing the incoming customer calls

C. Prioritizing the incoming customer calls

D. Escalating the incidents to more specialized personnel

36. Common calls from users for information or changes related to an IT service, such as a password reset, and does not require an RFC is known as what?

A. Service Request

B. Event

C. Alert

D. Problem

37. Daniels Manufacturing has a set of formal policies and processes that dictate how its various departments are to operate. They have also documented the roles, responsibilities, and authority of its staff. Together these provide guidance for the management of the business. The development, implementation, and use of this framework is called:

A. Service level agreement

B. Operational level agreement

C. Governance

D. Role-based management

38. In ITIL, what is the term used to describe the entity which is responsible for the delivery of a service?

A. Service Provider

B. Customer

C. Market Space

D. Service Lifecycle

39. In the past 6 months, Retro LLC has recorded over 1000 incidents for their online music service RetrollaOnline.com. Which of the following sources below may have detected and reported an incident?

I. An end user of the service
II. A Service Desk agent
III. A monitoring system

IV. Personnel from another IT department

A. I and II

B. I and III

C. I, III, and IV

D. All of these responses / All of the above

40. Your company is planning the future release of a service which will be offered to customers. Where are details regarding the planned service documented?

I. Service Pipeline
II. Service Catalogue
III. Service Portfolio
IV. Definitive Media Library

A. I and II

B. II and III

C. I and III

D. III and IV

ITIL Foundation
Practice Exam 3
Answer Key and Explanations

1. A - Suppliers are third-party organizations that provide essential goods or services to support another organization's service offering. In this case, Diamond Data Systems supplies services for the infrastructure resources that make up Markham's network. [ITIL - Selected Roles]

2. B - Under ITIL, a customer can include the business itself which provides the service. As an example, an Accounting department may require a service provided by the IT department, and is charged for it (directly or indirectly). (ITIL Service Strategy) - Service Management as a practice [ITIL - Service Management as a Practice]

3. B - Although utilized in a number of ITIL Core volumes, Service Level Management is most widely focused on in Service Design. (Service Design) - Selected Processes [ITIL - Selected Processes]

4. B - Resources are tangible or consumable assets such as IT infrastructure used to provide a service. In this scenario, the storage array is a resource providing disc space to coblus.com customers. [ITIL - Generic Concepts and Definitions]

5. B - Customer assets acts as the recipient of value provided by service assets, so statement B is false. (ITIL Service Strategy) - Service Management as a practice [ITIL - Service Management as a Practice]

6. C - Service Design covers the design principles and methods for converting strategic objectives in to portfolios of services and service assets. (Service Lifecycle) [ITIL - Service Lifecycle]

7. C - The number of Incidents due to Capacity Shortages would not be tracked by Service Level Management; rather it would be tracked by Capacity Management. (Service Design) - Selected Processes

8. A - The primary activity of the service desk is to act as the first point of contact to the customer. Problem Management is tasked with investigating the root cause of service disruptions. (Service Operation) - Selected Functions [ITIL - Selected Functions]

9. D - Service Transition provides guidance on managing the implementation, or transition, of a new or changed service into

operations, and draws heavily from Release Management, Program Management, and Risk Management. (Service Lifecycle) [ITIL - Service Lifecycle]

10. B - ITIL describes a Function as "A logical concept referring to people and automated measures that execute a defined process, an activity, or a combination thereof". (ITIL Service Design) [ITIL - Service Management as a Practice]

11. A - Capabilities are assets used to control or coordinate service resources. Generally speaking, capabilities are experience-based intangibles that are developed over time, such as teams, processes, or knowledge about a service. [ITIL - Generic Concepts and Definitions]

12. D - This concept is best represented by the term: Security. (Service Management) - Technology and Architecture [ITIL - Technology and Architecture]

13. C - The Core volumes of ITIL are Service Strategy, Service Design, Service Transition, Service Operation, and Continual Service Improvement (Service Lifecycle) [ITIL - Service Lifecycle]

14. A - Of the options listed, the Service Desk would report on the number of calls handled in total and by workstation. Resolved problems relate to Problem Management, and changes relate to Change Management. (Service Operation) - Selected Functions [ITIL - Selected Functions]

15. B - The Service Pipeline consists entirely of services in development. The Service Portfolio consists of services in development, services currently being offered, and services which have been retired. The Service Catalogue consists of services currently being offered. (Service Lifecycle) - Generic Concepts and Definitions [ITIL - Generic Concepts and Definitions]

16. C - Availability Management is comprised of the following four key aspects: Availability, Reliability, Maintainability, and Serviceability. (Service Design) - Selected Processes [ITIL - Selected Processes]

17. C - The Business Service Catalogue represents the customer's view and includes relationships to the business units and business processes which support IT services. (Service Lifecycle) - Generic Concepts and Definitions [ITIL - Generic Concepts and Definitions]

18. A - The Service Catalog is a document or database containing all IT services that are live or are available for deployment. It includes both customer facing services as well as supporting services required to deliver the service. The Service Catalog is a component of the Service Portfolio. [ITIL - Generic Concepts and Definitions]

19. C - Processes are not strategic; the providing organization or the customer must know what needs to be achieved. (ITIL Service Design) - Service Management as a practice [ITIL - Service Management as a Practice]

20. B - The Core volumes of ITIL are Service Strategy, Service Design, Service Transition, Service Operation, and Continual Service Improvement (Service Lifecycle) [ITIL - Service Lifecycle]

21. D - The Service Continuity Manager performs the activities above, and also develops and maintains a Service Continuity plan. (Service Design) - Selected Roles [ITIL - Selected Roles]

22. C - A strength of Service Automation is that it can operate outside of normal business hours. (Service Management) - Technology and Architecture [ITIL - Technology and Architecture]

23. D - A CAB, or Change Advisory Board, would be a key concept of Change Management. (Service Transition) - Selected Processes [ITIL - Selected Processes]

24. C - Of the items listed, identifying the root cause and creating a workaround is not the role of the Service Desk. (Service Operation) - Selected Functions [ITIL - Selected Functions]

25. C - The Change Manager is responsible for receiving, logging, and prioritizing RFCs; and will circulate them for review to members of the Change Advisory Board (CAB). (Service Transition) - Selected Roles [ITIL - Selected Roles]

26. A - The Process Model enables understanding and assists in articulating distinct process features, and is applicable to all processes. (ITIL Service Design) - Service Management as a practice [ITIL - Service Management as a Practice]

27. D - All of the reasons listed above are valid objectives of Service Measurement. (Continual Service Improvement) - Key Principles and Models [ITIL - Key Principles and Models]

28. A - Technology Metrics are typically associated with system components, application performance, and availability. (Continual Service Improvement) - Key Principles and Models [ITIL - Key Principles and Models]

29. A - A Release Unit describes the portion of a service or IT infrastructure that is typically released together, according to the organization’s release policy. Depending on the types or items of service asset or service component such as software and hardware, the unit may vary. [ITIL - Generic Concepts and Definitions]

30. A - Service Requests are a type of incident which do not involve a failure in the IT infrastructure. (Service Operation) - Selected Functions [ITIL - Selected Functions]

31. D - Problem Management would be responsible for the Time to resolve underlying root cause and prevent recurring incidents, not Incident Management. (Service Operation) - Selected Processes [ITIL - Selected Processes]

32. D - The Event Manager is responsible for the activities above, and is typically not identified as a dedicated role; rather, it is played by more than one resource. (Service Operation) - Selected Roles [ITIL - Selected Roles]

33. C - An Event is a notification created by service, CI, or monitoring tool; not a user. (Service Operation) - Generic Concepts and Definitions [ITIL - Generic Concepts and Definitions]

34. A - Demand Management best refelects the process described. Capacity Management is a process owned by Service Design. (Service Strategy) - Selected Processes [ITIL - Selected Processes]

35. A - Problem Management is responsible for identifying the cause of failures, not the Service Desk. (Service Operation) - Selected Functions [ITIL - Selected Functions]

36. A - In ITIL, a Service Request are common requests made by users for changes or information which do not require a Request for Change (RFC). (Service Operation) - Generic Concepts and Definitions [ITIL - Generic Concepts and Definitions]

37. C - Governance is the process of defining expectations, granting power, and verifying the performance of an organization or business. There are 3 primary areas of governance:

Enterprise, Corporate, and IT. [ITIL - Generic Concepts and Definitions]

38. A - A Service Provider is used to describe the entity which is responsible for the delivery of a service. (Service Design) - Generic Concepts and Definitions [ITIL - Generic Concepts and Definitions]

39. D - All response options listed above may have acted as a source for detecting and reporting an incident. Users detect and report incidents to the Service Desk. Service Desk agents can detect and directly record incidents. Monitoring systems are capable of detecting critical events and reporting them to the incident recording system. Personnel from another IT department may detect an incident and report it directly to the Incident recording system, or to the Service Desk. (Service Operation) - Generic Concepts and Definitions [ITIL - Generic Concepts and Definitions]

40. C - Details on planned services will be documented in the Service Pipeline; The Service Portfolio will contain information on planned, existing, and retired services, hence it will also contain details on the planned service. (Service Transition) - Generic Concepts and Definitions [ITIL - Generic Concepts and Definitions]

Knowledge Area Quiz: Selected Roles Practice Questions

Test Name: Knowledge Area Quiz: Selected Roles
Total Questions: 10
Correct Answers Needed to Pass: 7 (70.00%)
Time Allowed: 10 Minutes

Test Description

This practice test focuses on selected ITIL roles.

Test Questions

1. What role is responsible for documenting and publicizing processes, defining and reviewing KPIs for a given process, and contributing input to an ongoing Service Improvement Plan?

A. Service Owner

B. Process Owner

C. Change Manager

D. Service Level Manager

2. What are the four main roles of the RACI model?

A. Revenue, Accounting, Control, Investments

B. Plan, Do, Check, Act

C. Review, Audit, Confirm, Implement

D. Responsible, Accountable, Consulted, Informed

3. Brian has the ultimate responsibility for ensuring 3 specific processes are fit for purpose for the entire lifecycle of the service, from initial design through continual improvement. What is Brian's role?

A. A. Service manager

B. B. Process owner

C. C. Product manager

D. D. Process practitioner

4. Nora is responsible for the operational management of a process. What is Nora's role?

A. A. Process owner

B. B. Process practitioner

C. C. Process manager

D. D. Process planner

5. What role is responsible for the creation and management of Demand incentive and penalty schemes, monitors overall demand and capacity, and participates in the creation of SLAs?

A. IT Financial Manager

B. Service Level Manager

C. Demand Manager

D. Product Manager

6. What role is responsible for managing and maintaining the business and technical Service Catalogue while ensuring all information is consistent with that of the Service Portfolio?

A. Product Manager

B. Service Level Manager

C. Demand Manager

D. Service Catalogue Manager

7. What role is responsible for ensuring underpinning contracts are in line with the overall business, performs risk assessments of vendor agreements to ensure obligations are being met, and maintains a process for dealing with vendor disputes?

A. Supplier Manager

B. IT Financial Manager

C. Service Level Manager

D. Service Continuity Manager

8. What does the CAB acronym represent in ITIL?

A. Change Advisory Board

B. Configuration Access Board

C. Continuity Access Breach

D. Change About Business

9. What role is responsible for managing the work of 1st and 2nd line support staff, and managing Major Incidents as they occur?

A. Incident Manager

B. Problem Manager

C. Alert Manager

D. Event Manager

10. What role is considered to overlap between Security Management and Availability Management; and utilizes the Service Desk as an initial filter to perform tasks such as the checking of authority for user requests to access specific information areas?

A. Incident Manager

B. Alert Manager

C. Security Manager

D. Access Manager

Knowledge Area Quiz: Selected Roles Answer Key and Explanations

1. B - The Process Owner is responsible for the activities listed above; In addition, the Process Owner must address process issues, and ensure sufficient staff exists to carry out specific processes. (Service Management / Service Lifecycle) - Selected Roles [ITIL - Selected Roles]

2. D - The four main roles of the RACI model are Responsible, Accountable, Consulted, Informed. (Service Management / Service Lifecycle) - Selected Roles [ITIL - Selected Roles]

3. B - The process owner bears the responsibility and accountability for ensuring that a process is fit for purpose for its entire lifecycle. [ITIL Selected Roles]

4. C - The process manager is responsible for the operational management of a process, including coordination of the activities required to carry out, monitor, and report on the process. [ITIL Selected Roles]

5. C - The Demand Manager is responsible for the creation and management of Demand incentive and penalty schemes, monitors overall demand and capacity, and participates in the creation of SLAs. (Service Strategy) - Selected Roles [ITIL - Selected Roles]

6. D - The Service Catalogue Manager performs the activities above, and also agrees to and documents services, and ensures that information is in alignment with business processes. (Service Design) - Selected Roles [ITIL - Selected Roles]

7. A - The Supplier Manager performs the activities above, and also maintains a Supplier and Contracts Database (SCD). (Service Design) - Selected Roles [ITIL - Selected Roles]

8. A - The CAB stands for Change Advisory Board, and meets to review RFCs (Request for Changes) which have been raised. The CAB may be made up of different members, depending on the changes proposed. (Service Transition) - Selected Roles [ITIL - Selected Roles]

9. A - The Incident Manager is responsible for the activities above, and also maintains all relevant Incident Management systems. (Service Operation) - Selected Roles [ITIL - Selected Roles]

10. D - The Access Manager is responsible for managing access policies and procedures which are practiced by a number of different processes. (Service Operation) - Selected Roles [ITIL - Selected Roles]

Knowledge Area Quiz: Technology and Architecture Practice Questions

Test Name: Knowledge Area Quiz: Technology and Architecture
Total Questions: 10
Correct Answers Needed to Pass: 7 (70.00%)
Time Allowed: 10 Minutes

Test Description

This practice test focuses on the Technology and Architecture of Service Management.

Test Questions

1. What should a Service Management tool, which is implemented by an organization, always reference?

   A. The Service Catalogue

   B. The Service Pipeline

   C. The Service Portofolio

   D. The Configuration Managament System (CMS)

2. What type of technology is used to assist with License Management, and typically requires audit tools which can be run in an automated fashion from any network location; and is also capable of querying and receiving information on all CIs within the IT infrastructure?

   A. Alerting tools

   B. Remote control tools

   C. Discovery, Deployment, and Licensing Technologies

   D. Integrated Configuration Management System (CMS)

3. What type of reporting allows for a summary view of overall IT performance and availability, may provide real time information, and is often related in management reports to customers and users?

   A. Service Catalogue

   B. Discovery, Deployment, and Licensing Technologies

   C. Dashboards

   D. Alerts

4. What term best reflects a Service Management product's ability to be substantially enlarged, either via the amount of data it stores, or the via the amount of users it supports?

   A. Capacity

   B. Scalability

   C. Continuity

   D. Security

5. What term best reflects a Service Management product's ability to recover from a failure?

   A. Capacity

   B. Scalability

   C. Continuity

   D. Security

6. Which of the following service assets does Service Automation impact the performance of?

   I. Management
   II. People
   III. Process
   IV. Knowledge

   A. I and II

   B. I and IV

   C. II, III and IV

   D. All of these responses / All of the above

7. When applied properly, which of the following items are expected to be reduced by Service Automation?
   I. Costs
   II. Quality
   III. Warranty
   IV. Risks

   A. I and II

   B. II and III

   C. III and IV

   D. I and IV

8. Which of the following statements about Service Management tools are true?

   I. Service Management tools must always reference the Service Portfolio
   II. Processes should be modified to fit the tools

A. I and II

B. I

C. II

D. Both statements are false

9. The success of a Service Management tool is least likely to be dependent on which of the following items?

   A. Process

   B. People

   C. Perception

   D. Function

10. What is the term used to represent the coordination of business-related IT with the processes and approaches of IT Service Management?

   A. Business Service Management

   B. IT Continuity Service Management

   C. Service Automation

   D. Availability Management

Knowledge Area Quiz: Technology and Architecture Answer Key and Explanations

1. C - A Service Management tool which is implemented by an orgnanization should always reference the Service Portfolio. (Service Management) - Technology and Architecture [ITIL - Technology and Architecture]

2. D - An Integrated Configuration Management System (CMS) holds together any and all relevant CIs, along with their related attributes, in a centralized location; and is capable of linking to Incident, Problem, and Change records. (Service Management) - Technology and Architecture [ITIL - Technology and Architecture]

3. C - Dashboards provide a summary view of overall IT performance and availability, may provide real time information, and are often related in management reports to customers and users. (Service Management) - Technology and Architecture [ITIL - Technology and Architecture]

4. B - This concept is best represented by the term: Scalability. (Service Management) - Technology and Architecture [ITIL - Technology and Architecture]

5. C - This concept is best represented by the term: Continuity. (Service Management) - Technology and Architecture [ITIL - Technology and Architecture]

6. D - All of the items listed are significantly impacted by Service Automation. (Service Management) - Technology and Architecture [ITIL - Technology and Architecture]

7. D - When applied properly, costs and risks are expected to be reduced by Service Automation. (Service Management) - Technology and Architecture [ITIL - Technology and Architecture]

8. B - In general, processes should not be modified to fit the tools; rather, tools should be modified to fit the process. Service Management tools should always reference the Service Portfolio. (Service Management) - Technology and Architecture [ITIL - Technology and Architecture]

9. C - The success of a Service Management tool is primarily dependent upon Process, Function, and People. (Service Management) - Technology and Architecture [ITIL - Technology and Architecture]

10. A - Business Service Management represents the coordination between IT Service Management and business-related IT management. (Service Management) - Technology and Architecture [ITIL - Technology and Architecture]

ITIL Foundation Practice Exam 4 Practice Questions

Test Name: ITIL Foundation Practice Exam 4
Total Questions: 40
Correct Answers Needed to Pass: 30 (75.00%)
Time Allowed: 60 Minutes

Test Description

This is the fourth cumulative ITIL Foundation test which can be used as an indicator for overall performance. This practice test includes questions from all ITIL areas.

Test Questions

1. Which of the statements below concerning Value Creation are true?

   I. An organization uses Service Assets to create value in the form of goods and services.
   II. An organization relies on Capabilities to coordinate, control, and deploy Resources to produce value.

   A. I and II

   B. I

   C. II

   D. None of the above are true.

2. What ITIL Function acts as a single point of contact for IT users on a day to day basis?

   A. Incident Management

   B. Application Management

   C. Service Desk

   D. IT Operations Management

3. Justine is developing documentation that includes a cost-benefit analysis of a proposed new service, as well as ramifications of not implementing this new service. What is this type of documentation called?

   A. Risk Management

   B. Business Case

   C. TCO

   D. ROI

4. Fitness for use, or the assurance that products and services provided will meet certain specifications, reflects which concept below?

   A. Utility

   B. Resources

   C. Warranty

   D. Service Management

5. What ITIL Core volume is most closely aligned with the processes of Service Measurement, Service Improvement and Service Reporting?

   A. Continual Service Improvement

   B. Service Transition

   C. Service Design

   D. Service Strategy

6. Which of the items below holds together any and all relevant CIs, along with their related attributes, in a centralized location; and is capable of linking to Incident, Problem, and Change records?

   A. Definitive Media Store

   B. Discovery, Deployment, and Licensing Technologies

   C. Asset Management System (AMS)

   D. Integrated Configuration Management System (CMS)

7. What item below does not represent a primary step in developing a RACI chart?

   A. Identify the activities and processes

   B. Coordinate meetings to assign RACI codes

   C. Review SLA to ensure customer requirements are maintained

   D. Identify assignment gaps or overlaps

8. Which of the items below represents a service provider's investments across all market spaces, and includes the Service Catalogue, Service Pipeline, and Retired Services?

   A. Service Lifecycle

   B. Market Space

   C. Customer Assets

D. Service Portfolio

9. What ITIL function manages and maintains the IT infrastructure by managing the physical IT environment and provides Operations Control and Facilities Management?

A. Service Desk

B. Technical Management

C. Application Management

D. IT Operations Management

10. Which of the following ITIL Core volumes is applied when planning to design and modify services and processes?

A. Service Transition

B. Service Design

C. Continual Service Improvement

D. Service Strategy

11. Within the Service Lifecycle, how would the following events best be sequenced?

I. Service Design begins architecting the service
II. The service becomes part of the Service Catalogue
III. An organization makes a strategic decision to charter a service

A. I, II, III

B. III, I, II

C. II, III, I

D. III, II, I

12. The technical review board at a regional government agency is discussing how a recently discovered email virus could affect their operations, the probability of infection, and mechanisms for reducing the probability or impact. What is this process called?

A. Operations Management

B. Availability Management

C. Risk Management

D. Systems Management

13. What item is recorded in the SLA, and defines the expected times for which a customer should have access to a particular service?

A. Capacity

B. Demand

C. Utility

D. Availability

14. Which of the following processes are not associated with Service Design?

A. Service Catalogue Management

B. Service Level Management

C. Availability Management

D. Service Measurement

15. Which item below best reflects a group, team, or person that performs tasks relevant to a specific process?

A. Function

B. Best Practice

C. Role

D. Service

16. What are the two primary areas of service improvement to be expected from Service Automation?

I. Utility
II. Warranty
III. Security
IV. Capacity

A. I and II

B. II and III

C. III and IV

D. I and IV

17. Availability of services, controlling demand, and optimizing the use of existing capacity in a day-to-day service environment are most critical to which of the following ITIL Core volumes?

A. Service Design

B. Service Operation

C. Service Transition

D. Continual Service Improvement

18. A cloud-based email service provider guarantees in writing that its premier level customers will have access to live support staff 8x7x365. What is this type of guarantee called?

A. Operational level agreement

B. Service level agreement

C. Underpinning contract

D. Terms and conditions

19. A Service Desk structure consisting of multiple Service Desks which appear to form a single unit, can be located anywhere, and rely upon modern telecommunication technology is termed:

A. Centralized Service Desk

B. Off-shore Service Desk

C. Call Center

D. Virtual Service Desk

20. Under Service Design, what process is used by an organization to meet the objectives of confidentiality, information integrity, information availability, and authenticity of information?

A. Availability Management

B. Supplier Management

C. Service Level Management

D. Information Security Management

21. What role is responsible for ensuring agreed levels of service availability are maintained, monitors the actual availability of services achieved, and manages the Availability Plan?

A. Service Continuity Manager

B. Availability Manager

C. Capacity Manager

D. Demand Manager

22. Which of the following statements describing a process in ITIL are true?

I. A process includes all of the resources required to deliver the outputs.
II. A process may not define or revise organizational policies or standards

A. I and II

B. I

C. II

D. Both statements are false

23. What concept allows management to better understand a service's quality requirements, and presents both the associated costs and expected benefits?

A. Business Case

B. Technical Service Catalogue

C. Risk Analysis

D. Service Portfolio

24. What stage of the Deming Cycle requires the actual implementation of improvements to Service and Service Management processes?

A. Plan

B. Do

C. Check

D. Act

25. A broad concept reflecting an organization's informed decision making process, underpinned by data from the CMDB and CMS is known as what?

A. Service Catalogue

B. Quality Management System

C. Service Model

D. Service Knowledge Management System (SKMS)

26. To ensure that a process is continually improved and meets its objectives, it is most critical that the process has which of the following?

A. A Process Policy

B. Process Capabilities

C. Process Enablers

D. A Process Owner

27. Which of the following processes is not owned by the ITIL Core volume Service Transition?

A. Service Asset and Configuration Management

B. Change Management

C. Release and Deployment Management

D. Operation Management

28. Which of the Service Desk performance indicators below would least likely be measured by means of a customer survey?

A. Was the support call answered politely?

B. Were users provided guidance which was easy to follow and accurate?

C. Is the user perception of the company improved based on their experience with the Service Desk?

D. What was the percentage of calls answered within 45 seconds of entering the system?

29. Taking in to account the dominant pattern of the Service Lifecycle, what area focuses on policies, standards, and objectives; and creates outputs which are used by Service Design?

A. Service Transition

B. Service Design

C. Continual Service Improvement

D. Service Strategy

30. Which of the following items fall under the area of Process Control in the Process Model?

I. Process Owner
II. Process Resources
III. Process Policy
IV. Process Capabilities

A. I, III

B. II, IV

C. I, II, III

D. II, III, IV

31. Which of the following is not one of the three types of metrics under Continual Service Improvement (CSI)?

A. Technology Metrics

B. Process Metrics

C. Baseline Metrics

D. Service Metrics

32. Carpenter Consulting has been experiencing a spike in the level of support calls received over the past week. Many of the incidents reported require immediate attention to minimize disruptions to service. How should the priority of incidents be addressed?

A. Incidents should be prioritized based on the order each request was received.

B. Incidents should be prioritized based on the number of people affected for a given disruption.

C. Incidents should be prioritized based on their impact and urgency.

D. Incidents should be prioritized based on the importance/seniority of the person making the request.

33. What role is responsible for managing the CMS, as well as for managing the CIs and defining their naming conventions?

A. Service Asset Manager

B. Configuration Manager

C. Change Manager

D. Configuration Control Board

34. Which of the items below are valid performance indicators for the Service Desk?

I. Number of incidents resolved without escalation
II. Percentage of incidents properly logged
III. Number of telephone calls which were answered courteously, as reported in customer surveys

A. I and II

B. II and III

C. I and III

D. All of these responses / All of the above

35. Which of the following statements regarding Process Control are true?

I. Processes which are controlled can be repeated
II. The extent of control over a process should be defined after measurement metrics are determined for a given process
III. An organization should document and control defined processes

A. I and II

B. II and III

C. I and III

D. All of these responses / All of the above

36. A Problem whose root cause has already been determined, and for which a Workaround has been documented, is known as what?

A. Event

B. Known Error

C. Alert

D. Incident

37. The failure of a Configuration Item (CI), which has not yet impacted service, is known as what?

A. Incident

B. Event

C. Alert

D. Problem

38. A set of connected behaviors or actions which are performed by a person, team, or group in a specific context is known as what?

A. An Event

B. A Resource

C. A Service

D. A Role

39. What does Service Design pass on to Service Transition for each new service, major change to an existing service, or removal of a service?

A. Delta Release

B. Request for Change (RFC)

C. Incident

D. Service Design Package (SDP)

40. WebSphere Hosting, a web site hosting company, received a support request from one of its clients. The client requested that a backup of their web site's database be made available for download. A look at the SLA shows that this request should be accomodated by WebSphere Hosting within 24 hours. How best can this request be categorized?

A. A Known Error

B. A Service Level Requirement

C. A Change Request

D. A Service Request

ITIL Foundation
Practice Exam 4
Answer Key and Explanations

1. A - Both of the statements are true. (Service Strategy) - Key Principles and Models [ITIL - Key Principles and Models]

2. C - The Service Desk is the function which acts as a single point of contact for IT users on a day to day basis. (Service Operation) - Selected Functions [ITIL - Selected Functions]

3. B - A business case provides the justification for the implementation of a new service including the costs, benefits, and risks associated with the new service. [ITIL Generic Concepts and Definitions]

4. C - Warranty reflects "fitness for use" and assures customers that certain specifications are met for products or services; also, performance variation is reduced with warranty . (Service Management) - Generic Concepts and Definitions [ITIL - Generic Concepts and Definitions]

5. A - Continual Service Improvement is most closely aligned with the processes of Service Measurement, Service Improvement and Service Reporting (Service Lifecycle) [ITIL - Service Lifecycle]

6. D - An Integrated Configuration Management System (CMS) holds together any and all relevant CIs, along with their related attributes, in a centralized location; and is capable of linking to Incident, Problem, and Change records. (Service Management) - Technology and Architecture [ITIL - Technology and Architecture]

7. C - Reviewing SLAs does not represent a primary step in the development of a RACI chart. (Service Management / Service Lifecycle) - Selected Roles [ITIL - Selected Roles]

8. D - The Service Portfolio encompasses the Service Catalogue, Service Pipeline, and Retired Services. (Service Strategy) - Selected Processes [ITIL - Selected Processes]

9. D - The IT Operations Management function maintains the IT infrastructure by managing the physical IT environment and provides Operations Control and Facilities Management. (Service Operation) - Selected Functions [ITIL - Selected Functions]

10. B - Service Design provides guidance when designing and developing

services and processes. (Service Lifecycle) [ITIL - Service Lifecycle]

11. B - The proper sequence is for a decision to be made to charter a service, the service gets designed, the service ultimately becomes part of the organization's Service Catalogue. (Service Design) - Key Principles and Models [ITIL - Key Principles and Models]

12. C - Risk Management is the process of identifying, assessing, and controlling risks such as viruses, fire, or critical system failure. [ITIL Generic Concepts and Definitions]

13. D - Availability defines when a customer should expect to have access to a particular service and is defined in the SLA. (Service Design) - Generic Concepts and Definitions [ITIL - Generic Concepts and Definitions]

14. D - Service Measurement is a process covered by Continual Service Management. All of the others belong to Service Design. (Service Design) - Selected Processes [ITIL - Selected Processes]

15. C - Under ITIL, this statement best describes a Role. (ITIL Service Strategy) - Service Management as a practice [ITIL - Service Management as a Practice]

16. A - Utilitiy and Warranty are expected to be improved in services as a result of Service Automation. (Service Management) - Technology and Architecture [ITIL - Technology and Architecture]

17. B - Service Operation provides guidance on the effectiveness and efficiency for the delivery and support of operational (day-to-day) services to customers, and focuses heavily on the availability of services, controlling demand, and optimizing the use of existing capacity. (Service Lifecycle) [ITIL - Service Lifecycle]

18. B - Service level agreements (SLAs) are formally documented guarantees between a service provider and its customer that services will meet targets such as availability, reliability, and/ or capacity. SLAs may also define the roles and responsibilities of the customer or the service provider. SLAs may be extended to internal or external customers. [ITIL - Generic Concepts and Definitions]

19. D - A virtual Service Desk appears to users as a single support unit, even though it may consist of many personnel which are remotely dispersed. (Service Operation) - Selected Functions [ITIL - Selected Functions]

20. D - Information Security Management supports the main objectives of confidentiality, integrity, availability, and authenticity. (Service Design) - Selected Processes [ITIL - Selected Processes]

21. B - The Availability Manager performs the activities above, and also assesses the potential impacts of change in relation to service availability; may also attend the CAB. (Service Design) - Selected Roles [ITIL - Selected Roles]

22. B - Under ITIL, a process may indeed define and revise organizational policies or standards, as well as guidelines, activities, and work instructions if needed. (ITIL Service Strategy) - Service Management as a practice [ITIL - Service Management as a Practice]

23. A - A Business Case presents management with a service's quality requirements and associated delivery costs, in addition to models which outline what a service is expected to achieve. (Service Lifecycle) - Generic Concepts and Definitions [ITIL - Generic Concepts and Definitions]

24. D - The "Act" stage of the Deming Cycle requires implementation of the improvement to the Service and Service Management processes. (Continual Service Improvement) - Key Principles and Models [ITIL - Key Principles and Models]

25. D - The Service Knowledge Management System (SKMS) leverages underpinning data stored in the CMDB, which flows through the CMS, to make informad decisions. (Service Lifecycle) - Generic Concepts and Definitions [ITIL - Generic Concepts and Definitions]

26. D - The Process Owner is responsible for ensuring a process meets its objectives and is continually improved. (ITIL Service Design) [ITIL - Service Management as a Practice]

27. D - Operation Management is owned by the ITIL Core volume "Service Operation". (Service Transition) - Selected Processes [ITIL - Selected Processes]

28. D - The percentage of calls answered within X amount of seconds is a quantitative performance indicator not normally measured by a customer survey, and is typically tracked in an automated fashion. (Service Operation) - Selected Functions [ITIL - Selected Functions]

29. D - The dominant pattern of progress in the Service Lifecycle is for Service Design to follow Service Strategy;

Service Strategy focuses on the creation of policies, standards, and objectives (Service Lifecycle) [ITIL - Service Lifecycle]

30. A - The Process Owner and Process Policy fall under the area of Process Control in the Process Model; Process Resources and Process Capabilities fall under the area of Process Enablers.. (ITIL Service Design) - Service Management as a practice [ITIL - Service Management as a Practice]

31. C - The three types of metrics covered under Continual Service Improvement are Technology, Process, and Service metrics. (Continual Service Improvement) - Key Principles and Models [ITIL - Key Principles and Models]

32. C - When several incidents are being dealt with at the same time, priority must be determined based on the impact and urgency of each incident. (Service Operation) - Generic Concepts and Definitions [ITIL - Generic Concepts and Definitions]

33. B - The Configuration Manager is responsible for managing the Configuration Management System (CMS), as well as for managing the Configuration Items (CIs) and defining their naming conventions. (Service Transition) - Selected Roles [ITIL - Selected Roles]

34. D - All of the responses are valid performance indicators for the Service Desk. Some performance indicators are more qualitative and best measured via customer surveys, such as whether telephone calls were courteously answered. (Service Operation) - Selected Functions [ITIL - Selected Functions]

35. C - Under the Process Control activity, the extent of control over a process should be defined before measurement metrics are defined. (ITIL Service Design) - Service Management as a practice [ITIL - Service Management as a Practice]

36. B - A Known Error is a problem whose cause is known and for which a workaround has been documented. (Service Operation) - Generic Concepts and Definitions [ITIL - Generic Concepts and Definitions]

37. A - Even if the service has not yet been impacted, the failure of a CI is considered an incident. (Service Operation) - Selected Functions [ITIL - Selected Functions]

38. D - This statement represents the definition of a role. (Service Strategy) - Selected Roles [ITIL - Selected Roles]

39. D - A Service Design Package (SDP) is passed on from Service Design to Service Transition each time a service is added, changed, or removed. (Service Design) - Generic Concepts and Definitions [ITIL - Generic Concepts and Definitions]

40. D - ITIL defines standard services which are agreed to in the SLA as Service Requests. Service requests are handled using the Incident Management process. (Service Operation) - Generic Concepts and Definitions [ITIL - Generic Concepts and Definitions]

Knowledge Area Quiz: More Key Principles and Models Practice Questions

Test Name: Knowledge Area Quiz: More Key Principles and Models
Total Questions: 10
Correct Answers Needed to Pass: 7 (70.00%)
Time Allowed: 10 Minutes

Test Description

This practice test targets additional Key Principles and Models of ITIL.

Test Questions

1. The Continual Service Improvement phase consists of:

   A. Developing appropriate IT services, including architecture, processes, policy, and documents.

   B. Achieving effectiveness and efficiency in providing and supporting services in order to ensure value for the customer and the service provider.

   C. Designing, developing, and implementing service management as a strategic resource.

   D. Creating and maintaining customer value through design improvement, service introduction, and operation

2. What is the most important objective of Service Design?

   A. Development of policies and standards

   B. Accuracy of the SLA.

   C. The design of new or modified services for introduction into a production environment.

   D. Competing effectively with other organizations

3. Reviewing progress, fulfillment, effectiveness, and efficiency are key components of what type of review?

   A. Due care

   B. Process design

   C. Process alignment

   D. Process measurement

4. While aligning IT services with business needs and goals, the service management team at Fairfax Supermarkets reviews the four perspectives of IT service management as they apply to their company. What are these perspectives?

   A. People, processes, partners, products

   B. Teams, infrastructure, processes, policies

   C. Priorities, budget, resources, revenue

   D. Technology, architecture, business practices, customer requirements

5. What is used to measure how well an organization is doing in its stated goal of reducing calls to the help desk for password resets by implementing a self service tool on the corporate intranet?

   A. Key performance indicators (KPIs)

   B. Critical success factors (CSFs)

   C. Operational level agreements (OLAs)

   D. Patterns of business activities (PBAs)

6. Which of the following techniques is used for root cause analysis?

   A. Rummler-Brache swim-lane diagram

   B. Ishikawa diagram

   C. Kano Model

   D. Pareto diagram

7. International Products and Manufacturing, LTD is a large conglomerate with many business units. The organization has decided to spin-off its IT and finance functions into a single, separate business unit that will support all other business units. What type of service provider is this new unit?

   A. Consolidated Unit

   B. Corporate Operations Unit

   C. Business Services Unit

   D. Shared Services Unit

8. According to the Deming quality circle, a number of steps must be performed repeatedly in order to ensure good performance. Which of the following are the correct sequence of steps?

   A. Act-Check-Do-Plan

   B. Check-Plan-Act-Do

   C. Do-Plan-Check-Act

   D. Plan- Do-Check-Act

9. An SLA that governs the availability of a specific application during specific hours for all clients is an example of what type of SLA?

   A. Technical SLA

   B. Client-based SLA

   C. Catalog-based SLA

   D. Service- based SLA

10. The two types of assets are ________________ and ______________.

    A. Quantifiable and qualitative

    B. Skills and manufacturing capacity

    C. Goods and services

    D. Resources and capabilities

Knowledge Area Quiz: More Key Principles and Models Answer Key and Explanations

1. D - Continual service improvement is the phase of creating and maintaining customer value through design improvement, service introduction, and operation. [ITIL - Key Principles and Models]

2. C - According to ITIL, the most important objective of Service Design is the design of new or modified services for introduction into a production environment. [ITIL - Key Principles and Models]

3. D - In order to lead and manage the development process effectively, regular assessments must be performed. There are four elements that can be investigated: progress, fulfillment, effectiveness, and efficiency of the process. [ITIL - Key Principles and Models]

4. A - People, processes, partners, and products are the four perspectives of IT service management. These are collectively known as the 4P's. [ITIL Key Principles and Models]

5. A - Key performance indicators (KPIs) are used to quantify and measure service elements that contribute to meeting an organization's objectives. In this scenario, reducing the volume of password reset calls is a critical success factor (CSF) that could be measured by a KPI that reports the number and type of calls over time. [ITIL - Key Principles and Models]

6. B - The Ishikawa Diagram is a technique that helps a team identify all possible causes of a problem. [ITIL - Key Principles and Models]

7. D - Functions such as finance, IT, and logistics are not always at the core of an organization's competitive advantage, and need not be maintained at the corporate level. The services of such shared functions are often consolidated into an autonomous unit called a shared services unit. [ITIL - Key Principles and Models]

8. D - Deming developed a step-by-step improvement approach called the Plan-do-Check-Act Cycle (P-D-C-A). [ITIL - Key Principles and Models]

9. D - Service-based SLAs cover a service for all clients. [ITIL - Key Principles and Models]

10. D - Resources and capabilities are types of assets. Organizations use them to create value in the form of

goods and services. [ITIL - Key Principles and Models]

ITIL Foundation Practice Exam 5 Practice Questions

Test Name: ITIL Foundation Practice Exam 5
Total Questions: 40
Correct Answers Needed to Pass: 30 (75.00%)
Time Allowed: 60 Minutes

Test Description

This is the fifth cumulative ITIL Foundation test which can be used as an indicator for overall performance. This practice test includes questions from key ITIL areas.

Test Questions

1. Which of the following statements regarding an ITIL "process" is true?

   A. Processes are measurable because they are performance-oriented.

   B. Processes do not respond to a specific event.

   C. Processes are not measurable, because they support functions.

   D. A process is a specialized subdivision of an organization.

2. Which of the following represents a means of delivering value to customers by facilitating outcomes the customers want to achieve, without the ownership of specific costs or risks?

   A. A value

   B. An asset

   C. A system

   D. A service

3. A monitoring system at HostIT INC indicated that a recent change to the infrastructure resulted in the failure of five client servers. To plan the next course of action, HostIT needs to act immediately, and does not have time to coordinate a formal review of next steps. Under ITIL, what is the acronym of the group of people HostIT will assemble to review these highly urgent changes?

   A. ECAB

   B. CAB

   C. ACB

D. CAB-EC

4. Which of the following is not a process associated with Service Operation?

A. Access Management

B. Service Desk

C. Incident Management

D. Event Management

5. Delta Solutions offers a number of specialized organizational capabilities which provide value to their customers in the form of services. What term best reflects these capabilities?

A. Service Management

B. Encapsulation

C. ITIL

D. Agency Principle

6. HostIT INC, an outsourcing provider of website hosting and email management, guarantees 24 X 7 network availability and unlimited server storage for its customers. What term best reflects these guarantees?

A. Capacity

B. Utility

C. Warranty

D. Availability

7. What process is made up of nine basic activities, including: planning, building and testing, preparing the deployment, Early Life Support; and reviewing and closing?

A. Service Validation and Testing

B. Change Management

C. Event Management

D. Release and Deployment Management

8. The attributes of a service that have a positive effect on activity performance and represent fitness for purpose, is best reflected by what term?

A. Service

B. Value

C. Warranty

D. Utility

9. SmartIT INC ensures that deployment plans are formalized, release packages are successfully implemented, customers are aware of service updates, and service disruptions are minimized upon each new update to their web content publishing platform. What process is SmartIT INC following?

A. Change Management

B. Release and Deployment Management

C. Incident Management

D. Configuration Management

10. SST Logica keeps a database with details of all the services supplied to their customers. In addition, this database contains information on the Configuration Items (CIs) each service is reliant upon; items which are not visible to their customers. What is the term for this database?

A. Configuration Management System

B. Business Service Catalogue

C. Technical Service Catalogue

D. Service Portfolio

11. Select the best terms to fill in the blanks: The _______ is a subset of the ________ and only includes services which are approved and active in Service Operation.

A. Service Portfolio, Service Catalogue

B. Service Catalogue, Service Portfolio

C. Service Portfolio, Business Catalogue

D. Configuration Management System, Service Portfolio

12. Financial management, Demand management, and Service Portfolio management are three processes most critical to which phase of the Service Lifecycle?

A. Service Operation

B. Service Transition

C. Service Design

D. Service Strategy

13. The IT department of Western Manufacturing guarantees the corporate ERP system will be available between 12:00AM Monday-10:00PM Sunday. What is this type of guarantee called?

A. A. Operational level agreement

B. B. Service level agreement

C. C. Underpinning contract

D. D. Service Window

14. The ITIL Service Lifecycle consists of multiple phases. Which of the following items is not a phase of the Service Lifecycle?

A. Continual Service Improvement

B. Service Strategy

C. Service Design

D. IT Service Management

15. Which item does not reflect a component of Warranty, or fitness for use?

A. Continuity

B. Capacity

C. Utility

D. Availability

16. In order to derive meaningful metrics from Continual Service Improvement (CSI), what must be established as a result of the first measurement?

A. KPI

B. Baseline

C. CSF

D. Core value

17. Which of the following statements is false regarding Configuration Management?

A. A Configuration Management Database (CMDB) stores the attributes and relationships of Configuration Items (CIs) within the organization.

B. A Configuration Structure represents the relationships between all Configuration Items (CIs) within a given configuration.

C. Ensuring the proper management of Configuration Items (CIs) is the responsibility of the Configuration Control activity under Service Transition.

D. A Configuration Management Database (CMDB) may contain one or many Configuration Management Systems (CMSs)

18. A customer calls in to the Delta Solutions help desk to have his Intranet password reset. Since he had three failed login attempts, the Intranet system locked him out. What term best reflects this type of customer request?

A. Minor Incident

B. Error Call

C. Service Request

D. Request for Change (RFC)

19. TechCo, an IT service provider, is known for being a market leader in the field of complex system development. Which of the following items would not be considered an asset which TechCo possesses?

A. Infrastructure

B. Capital

C. Value

D. Knowledge

20. "Fitness for Use" is used to explain which of the following concepts?

A. Service

B. Warranty

C. Value

D. Utility

21. A cloud-based CRM service provider has technical support contracts on its database systems with the developer of the database software. The contracts guarantee 2-hour response time on all severity level 1 trouble tickets. What is this type of guarantee called?

A. Service level agreement

B. Underpinning contract

C. Operational level agreement

D. Third-party agreement

22. Jeanie uses her company's internet connection to do the research required for her job as a marketing manager. In this scenario, Jeanie is the:

A. Customer

B. User

C. Client

D. Internal customer

23. Select the best terms to fill in the blanks: The ______ is what the customer receives, and the _______ affirms how it will be delivered.

A. Value, Utility

B. Utility, Warranty

C. Value, Warranty

D. Warranty, Utility

24. SST Logica tracks all of the resources that are active in the various phases of the Service Lifecycle. The three main subsets SST Logica relies on are the Service Catalogue, Service Pipeline, and Retired Services. Together, what do these three service subsets represent?

A. Service Portfolio

B. Lines of Service

C. Service Lifecycle

D. Value

25. Defining the market, development of the offer, development strategic assets, and preparation for implementation are critical activities for which phase of the Service Lifecycle?

A. Service Transition

B. Service Strategy

C. Service Design

D. Service Operation

26. Which of the following items would not be a Key Performance Indicator (KPI) of Service Catalogue Management?

A. The percentage of Service Level Agreement infractions incurred by the performing organization

B. The amount of differences discovered between information stored in the Service Catalogue, and the actual state of services maintained by the organization

C. Number of incidents handled by the Service Desk using information stored in the Service Catalogue

D. The percentage of services which have been delivered, in relation to the total amount of services which exist in the Service Catalogue

27. Which of the following is not a technique used when seeking financing for ITIL projects?

A. Pre-Program ROI

B. Business case

C. Post-Program ROI

D. Value

28. What is the name of the formal plan to implement improvements to an IT Service, and is an output of the Continual Service Improvement (CSI) plan phase?

A. RACI

B. Release Plan

C. Service Improvement Plan (SIP)

D. Request for Change (RFC)

29. Which of the following items best reflects a decision-making, support, and planning instrument that prepares for the likely consequences of a business action?

A. Analytic model

B. Business case

C. Return on investment (ROI)

D. Simulation

30. What term is used to describe a significant change of state related to the management of an IT Service or Configuration Item?

A. Alert

B. RFC

C. Event

D. Red flag

31. Fitness for use, or the availability and reliability in continuity and security; representing a decline in possible losses, is best reflected by what term?

A. Service

B. Warranty

C. Value

D. Utility

32. What generic concept is critical for an IT service to succeed, and must be measured regularly by Key Performance Indicators (KPIs)?

A. Overhead

B. Critical Success Factor (CSF)

C. Core Value

D. Service Lifecycle

33. Delta Solutions is in the process of developing a new Internet Phone service which will allow customers to make telephone calls directly from their web browser. This new service will be available to customers in the European market only. Delta Solutions plans to release the service within the next six to twelve months. Where is this service most likely tracked by Delta Solutions?

A. Service Catalogue

B. Service Pipeline

C. Service Lifecycle

D. Retired Services

34. What process, most closely associated with Service Operation, is responsible for managing the lifecycle of all Service Requests.

A. Service Desk

B. Incident Management

C. Request Fulfillment

D. Service Portfolio Management

35. TechCo, an IT service provider, strives to focus heavily on where and how to compete in the market; distinguish its capabilities from its competitors; and view the services it provides as a strategic asset which must be constantly improved. What

Service Lifecycle phase best reflects these goals?

A. Service Design

B. Continual Service Improvement

C. Service Transition

D. Service Strategy

36. How quickly a service can be restored to service is a measure of which of the following?

A. Serviceability

B. Availability

C. Maintainability

D. Reliability

37. What term best represents a component of the Service Knowledge Management System where supplier contracts are managed throughout their lifecycle?

A. Supplier and Contract Database (SCD)

B. Service Catalogue

C. Service Portfolio

D. Supplier Catalogue

38. "Fitness for Purpose" is used to explain which of the following concepts?

A. Utility

B. Value

C. Service

D. Warranty

39. HostIT INC has defined a set of activities for its System Administrators to follow when provisioning server space for a new account. As such, HostIT is able to consistently describe what has to be done, what the expected results are, and how to measure the performance of the newly provisioned account. What is the ITIL term used to represent a structured set of activities designed to accomplish a defined objective?

A. Value

B. Procedure

C. Work Instruction

D. Process

40. Your organization has decided to outsource one of its services to a vendor. Which of the following is not considered a risk of outsourcing?

A. The outsourcing vendor can end up replacing your organization to the customer

B. The outsourcing vendor may need to compete with other vendors for your outsource business

C. Your organization may become dependent on the outsourcing organization

D. The vendor may damage your organization's reputation

ITIL Foundation
Practice Exam 5
Answer Key and Explanations

1. A - All processes are measurable, because they are performance-oriented. In addition, processes respond to specific events and have specific results. A "function" is a specialized subdivision of an organization. [ITIL - Service Management as a Practice]

2. D - A service represents a means of delivering value to customers by facilitating outcomes the customers want to achieve, without the ownership of specific costs or risks - Service Lifecycle (basic concept) [ITIL - Service Management as a Practice]

3. A - ECAB is the term used to describe the Emergency Change Advisory Board (ECAB), formally known as the CAB-EC in ITIL V2. The ECAB will most likely be assembled, as the change is critical and cannot wait for a formal Change Advisory Board (CAB) meeting. [ITIL - Selected Processes]

4. B - Although the Service Desk is associated with Service Operation, it is considered a function; not a process. [ITIL - Selected Processes]

5. A - The term Service Management is most applicable to the specialized organizational capabilities which provide value to customers in the form of services. Service Lifecycle (basic concept) [ITIL - Service Management as a Practice]

6. C - The guarantees of Capacity and Availability reflect the concept of Warranty. - Service Strategy [ITIL - Generic Concepts and Definitions]

7. D - These are basic process activities of Release and Deployment Management. [ITIL - Selected Processes]

8. D - Fitness for purpose describes the concept of "Utility" - Service Strategy [ITIL - Generic Concepts and Definitions]

9. B - Release and Deployment Management ensures that deployment plans are formalized, release packages are successfully implemented, customers are aware of service updates, and service disruptions are minimized. [ITIL - Selected Processes]

10. C - The Technical Service Catalogue encompasses information on services which is visible to the client (the Business Service Catalogue), as well as internal information regarding services; such as their relationship with CIs. -

Service Design [ITIL - Selected Processes]

11. B - The Service Catalogue is a subset of the Service Portfolio and only includes services which are approved and active in Service Operation. [ITIL - Selected Processes]

12. D - Financial management, Demand management, and Service Portfolio management are the three processes most critical to Service Strategy. [ITIL - Service Lifecycle]

13. A - Operational level agreements (OLAs) are formally documented guarantees between a service provider and another part of the same organization that services will meet targets such as availability, reliability, and/ or capacity. OLAs are similar to SLAs, but are only used between entities in the same organization. [ITIL - Generic Concepts and Definitions]

14. D - The five phases of the Service Lifecycle are Service Strategy, Service Design, Service Transition, Service Operation, and Continual Service Improvement. [ITIL - Service Lifecycle]

15. C - The primary concepts which represent Warranty are Availability, Capacity, Continuity, and Security. - Service Strategy [ITIL - Generic Concepts and Definitions]

16. B - A baseline must first be established in order to chart performance from future measurements. [ITIL - Key Principles and Models]

17. D - A Configuration Management System (CMS) may contain one or more Configuration Management Databases (CMDB), but a CMDB may not contain a CMS. All of the other statements are true. [ITIL - Selected Processes]

18. C - A Service Request best reflects this customer request, as it was not caused by an underlying fault in the IT infrastructure. [ITIL - Generic Concepts and Definitions]

19. C - Knowledge, capital, and infrastructure are all types of service assets which contribute to the basis of "value" for a service. - Service Strategy [ITIL - Service Lifecycle]

20. B - Fitness for use describes the concept of "Warranty" - Service Strategy [ITIL - Generic Concepts and Definitions]

21. B - Underpinning contracts (UCs) are formal agreements between a service provider and a third-party supplier that provides good or services essential for the delivery of an IT service to a customer. Similar to SLAs and OLAs, UCs define the targeted levels of service such as availability, reliability, and/ or capacity. [ITIL - Generic Concepts and Definitions]

22. B - Jeanie is a user in this scenario. ITIL defines a customer as a person that negotiates for and procures IT services, whereas a user is a person who uses an IT service on a daily basis. [ITIL - Generic Concepts and Definitions]

23. B - Utility is what a customer receives (fitness for purpose) and warrant affirms how it will be delivered (fitness for use) - Service Strategy [ITIL - Generic Concepts and Definitions]

24. A - An organization's Service Portfolio is comprised of the Service Catalogue, Service Pipeline, and Retired Services. - Service Strategy [ITIL - Generic Concepts and Definitions]

25. B - These four activities are critical to the Service Strategy phase. - Service Strategy [ITIL - Service Lifecycle]

26. A - The percentage of Service Level Agreement infractions incurred by the performing organization pertains to Service Level Management, not Service Catalogue Management. [ITIL - Key Principles and Models]

27. D - Business case, Pre-program ROI, and Post-program ROI are all techniques used when seeking financing for an ITIL project. - Service Strategy [ITIL - Service Lifecycle]

28. C - A Service Improvement Plan (SIP) is the formal plan used to implement improvements to an IT Service; and is a result of the CSI plan phase. [ITIL - Selected Processes]

29. B - This statement best describes a business case, often used to justify investments in service management.- Service Strategy [ITIL - Service Management as a Practice]

30. C - An event describes a significant change of state related to the management of an IT Service of Configuration Item. An alert is a notification created by an IT service for a specific event. [ITIL - Generic Concepts and Definitions]

31. B - Fitness for use describes the concept of "Warranty" - Service Strategy [ITIL - Generic Concepts and Definitions]

32. B - For an IT service to succeed, one or more Critical Success Factors (CSFs) must be identified and measured by associated Key Performance Indicators (KPIs). [ITIL - Generic Concepts and Definitions]

33. B - The Service Pipeline reflects services that are still in development for a specific customer or market. - Service Strategy. [ITIL - Generic Concepts and Definitions]

34. C - Request Fulfillment, under Service Operation, is responsible for managing the overall lifecycle of all Service Requests. [ITIL - Selected Processes]

35. D - Marketing focus, distinguishing one's capabilities, and performance anatomy are all goals of Service Strategy.- Service Strategy [ITIL - Service Lifecycle]

36. C - Maintainability is the measure of how quickly a service can be restored in the event of an outage. It is often reported as mean time to restore service (MTRS) or downtime. [ITIL - Generic Concepts and Definitions]

37. A - The SCD is used to manage the entire lifecycle of supplier contracts, and is a component of the Service Knowledge Management System. [ITIL - Selected Processes]

38. A - Fitness for service describes the concept of "Utility" - Service Strategy [ITIL - Generic Concepts and Definitions]

39. D - A process is defined as a structured set of activities designed to accomplish a defined objective. A procedure is a specified way to carry out a process. A work instruction defines how an activity within a procedure should be carried out. [ITIL - Generic Concepts and Definitions]

40. B - Substitution, disruption, and distinctiveness are all risks of outsourcing. Vendor competition is not one of them. - Service Strategy [ITIL - Service Management as a Practice]

Knowledge Area Quiz: More Selected Processes Practice Questions

Test Name: Knowledge Area Quiz: More Selected Processes
Total Questions: 12
Correct Answers Needed to Pass: 8 (66.67%)
Time Allowed: 15 Minutes

Test Description

This practice test focuses on additional processes of ITIL.

Test Questions

1. The objective of ____________ is to detect events, analyze them and determine the right management action.

   A. Exception management

   B. Support services

   C. Event management

   D. Incident management

2. A____________ is an unplanned interruption to an IT service, while a (an)____________ is a random occurrence that is important to management.

   A. Technical fault, management fault

   B. Event, incident

   C. Quantities issue, qualitative issue

   D. Incident, event

3. Which of the following is NOT a result of a service improvement?

   A. Value on Investment (VOI)

   B. Return on Investment (ROI)

   C. Service Level Agreement (SLA)

   D. Total Cost of Ownership (TCO)

4. Widgets International is rolling out a major upgrade to its ERP system. The Help Desk will be manned with extra staff for the first two weeks after the deployment. This is an example of ______________.

   A. Service operation

   B. Early Life Support

C. Transition planning

D. A workaround

5. Which of the following processes results in a workaround?

A. Systems management

B. Problem management

C. Event management

D. Incident management

6. Which of the following methods will assist Logistics, Ltd in identifying places to cut costs while maintaining service quality across the enterprise?

A. Value contribution

B. Service provisioning

C. Service Portfolio Management

D. Decision making framework

7. What is the benefit of a multi-level SLA?

A. It only covers subjects that are relevant to a specific service relating to a specific client.

B. All requirements are captured in a single document.

C. It is generic.

D. It keeps the SLAs under control and reduces the need for frequent updates.

8. Release types can be categorized into several groups. What type of release is typically implemented as a temporary workaround?

A. Minor release

B. Major release

C. Interim release

D. Emergency release

9. John is the leader of a committee that meets quarterly to determine if the existing IT infrastructure is sufficient to support current and predicted demand. What is this process called?

A. Availability planning

B. Demand management

C. Capacity planning

D. Business impact analysis

10. Data can be placed into one of three categories:

A. Operational, tactical, and strategic

B. Tactical, strategic, and competitive

C. Confidential, public, private

D. Operations, finance, and market

11. Service Operation is impossible without the existence of a demand that consumes the product. This is called:

A. Consumption cycle

B. Synchronous production and consumption

C. Conspicuous consumption

D. Law of Supply and Demand

12. Which of the following is a valid response to an event?

A. Event logging

B. Creation of an RFC

C. Human interaction

D. All of these responses / All of the above

Knowledge Area Quiz: More Selected Processes Answer Key and Explanations

1. C - The objective of event management is to detect events, analyze them, and determine the right management action. [ITIL - Selected Processes]

2. D - An incident in an unplanned interruption to an IT service, while an event is a random occurrence of importance to management. [ITIL - Selected Processes]

3. D - Key benefits / results of service improvements include:
   - **A.** Increased Return on Investment (ROI)
   - **B.** Increased Value on Investment (VOI)

   In addition, Service Level Agreements (SLAs) must be monitored to ensure service improvement targets are being met.
   [ITIL - Selected Processes]

4. B - Early Life Support is intended to offer extra support after the deployment of a new or changed service. [ITIL - Selected Processes]

5. D - Incident management is concerned with finding the quickest path to service restoration. The workarounds identified in this process address only the symptoms affecting the user experience. Problem management focuses on the identification of the root cause and development of a resolution to prevent the problem from recurring. [ITIL - Selected Processes]

6. C - Service Portfolio Management (SPM) is a dynamic method to govern investments in Service Management across the enterprise in terms of financial values. [ITIL - Selected Processes]

7. D - Multi-level SLAs combine corporate, client, and service level SLAs and diminishes the need for frequent updates. [ITIL - Selected Processes]

8. D - Emergency releases are usually implemented as a temporary solution for a problem or unknown error. [ITIL - Selected Processes]

9. C - Capacity planning is the process responsible for reviewing, analyzing, and understanding the current load capabilities of a service or its supporting infrastructure, and developing a path forward to ensure that anticipated demand can continue to be met. [ITIL - Selected Processes]

10. A - Data can be classified on 3 levels: operational, tactical, and strategic. [ITIL - Selected Processes]

11. B - Synchronous production and consumption is a pull system, in which consumption cycles stimulate the production cycles. [ITIL - Selected Processes]

12. D - The response to an event can have many forms, and a combination of response forms is possible. [ITIL - Selected Processes]

ITIL Foundation Practice Exam 6 Practice Questions

Test Name: ITIL Foundation Practice Exam 6
Total Questions: 40
Correct Answers Needed to Pass: 30 (75.00%)
Time Allowed: 60 Minutes

Test Description

This is the sixth cumulative ITIL Foundation test which can be used as an indicator for overall performance. This practice test includes questions from key ITIL areas.

Test Questions

1. The four Ps of Service Design are:

   A. Planning, Policy, Procedures, and Process

   B. Planning, Preparation, Processes, Procedures

   C. Personnel, Processes, Products, and Partners

   D. Personnel, Process, Policy, Procedure

2. Mean time between failures (MTBF) is an indication of which service property?

   A. Maintainability

   B. Serviceability

   C. Reliability

   D. Availability

3. Global Network Services has been contracted to provide maintenance services for a mid-sized manufacturing company. How well Global Network Services meets its contractual obligations is known as which of the following?

   A. Maintainability

   B. Availability

   C. Reliability

   D. Serviceability

4. Which of the following are not aspects of good service design?

   A. Systems

B. Design of Service Solution

C. Service Architecture

D. Processes

5. Connaught Construction conducts a quarterly management meeting to review how well current IT services are meeting the current needs of the business, as well as identifying potential new needs. This activity falls within the scope of which Lifecycle phase?

A. Service Design

B. Continual Service Improvement

C. Service Strategy

D. Service Operation

6. What is a consequence of implementing an overly complex process?

A. Service quality KPIs will increase.

B. Service quality can be lowered if processes are excessively cumbersome.

C. Relevant KPIs become invisible

D. Changes become easier to manage.

7. The time within which a function is back up after a failure is called:

A. Mean Time Between Failures

B. Mean Time to Restore Service

C. Service Recovery Objective

D. Restore Point Objective

8. Atlantic & Pacific Telecom has developed a set of criteria for use in balancing conflicting project and organization objectives. These criteria are called:

A. Strategy definitions

B. Tension metrics

C. Role definitions

D. Strategic objectives

9. Configuration items (CIs) are collected, stored, managed, updated, analyzed and reviewed in which of the following?

A. Service knowledge management system

B. Configuration records

C. Configuration management system

D. Configuration management database

10. A successful ITIL implementation requires support at which level of an organization?

A. Senior management

B. All levels

C. Line workers

D. Executive

11. Which of the following is not one of the three main types of metrics as defined by continual Service Improvement

A. Process

B. Service

C. Technology

D. Supplier

12. Resolving the conflict between maintaining standard service offerings and responding to changes to the environment in which the service is offered is part of the ________________ phase.

A. Service transition

B. Service operation

C. Service design

D. service strategy

13. Enterprise content management platform users asking for new features or changes to existing applications is usually the first phase of:

A. Service implementation

B. Technology implementation

C. Requirements gathering

D. The Service Design Phase

14. Which phase of the Service Lifecycle follows Service Transition?

A. Continual Service Improvement

B. Service Operation

C. Process Development

D. Service Launch

15. Fill in the blank: ________________________ is the core of all other phases in the Service Lifecycle.

A. Service Transition

B. Service Strategy

C. Service Operation

D. Service Design

16. The Service Lifecycle is a closed loop process. What phase is the input for the Service Design phase?

A. Service Development

B. Service Portfolio

C. Service Transition

D. Service Strategy

17. The phases of the Deming Cycle are:

A. Plan Prioritize Predict Prevent

B. Plan Do Check Act

C. Design Build Deploy Support

D. Design Build Test Release

18. What document defines and regulates the relationship between IT groups, including goods and services to be provided and the roles and responsibilities of each party?

A. Service Catalog

B. Operational Level Agreement

C. Service Time Agreement

D. Service Level Agreement

19. Organizational control and governance systems such as SOX or ISO have an impact on the design of ______________.

A. Shared services

B. Services

C. Measurement systems

D. Business organization

20. The SMART model is used to elaborate:

A. SLAs

B. Processes

C. Organizational goals

D. Requirements

21. Which of the following would be initiated with a service request?

A. Adding features to a website

B. Increasing the maximum attachment size for all email accounts

C. Replacing a failed drive

D. Installing a new printer for a single department

22. A baseline could be created for which of the following?

A. Software configurations

B. Uptime percentage

C. Bandwidth utilization

D. All of the above

23. How is service value calculated?

A. Service Utility + Service Warranty

B. Service Cost + Service Features

C. Service Uptime + Service Utilization

D. Service Utility + Service Cost

24. Customer needs are identified in which process?

A. Customer service management

B. Service Strategy

C. Service Design

D. Business relationship management

25. Which of the following is NOT a Service Design process?

A. Process management

B. Capacity management

C. Service Catalog management

D. Supplier management

26. Which of the following is a type of chart used to help monitor and report on achievement of Service Level Agreement Levels?

A. SWOT

B. SDLC

C. SLAM

D. SWAG

27. Which of the following tasks is NOT the responsibility of the CSI Manager?

A. Making knowledge management a permanent part of the daily routine

B. Taking care the proper monitoring tools are installed

C. Benchmarking

D. Defining monitoring demands together with the service level manager

28. Fill in the blank: ____________________is a specified way to carry out an activity.

A. Process

B. Procedure

C. Operating instruction

D. Task

29. Deploying a new web browser to all users in the finance department as part of a quarterly update package is an example of what type of change?

A. Version change

B. Technology change

C. Standard change

D. Regular change

30. A request for change is prioritized by________________ in collaboration with the initiator.

A. stakeholders

B. The service owner

C. The change manager

D. The change review board

31. Connecting Point has decided to bundle a set number of additional support hours beyond the basic support included with their Managed Hosting offering to stand out from their competitors' service offerings. This is called:

A. Delivery of Value through core services

B. Service Design.

C. The Formative Theorem

D. Developing a differentiated offering

32. The likelihood that a risk will occur and its possible impact determine ____________________.

A. The risk category of the change

B. The test plan

C. The recovery plan

D. The business impact of the change

33. Understanding why the volume of password related calls is down 25% over the last year is an example of which element of the DIKW structure?

A. Data

B. Information

C. Knowledge

D. Wisdom

34. Fill in the blank: ____________________ is a convenient tool for mapping out different configuration levels at which building and testing must take place.

A. V model

B. Rummler-Brache swim-lane diagram

C. Traceability matrix

D. RACI diagram

35. The DIKW structure is used for ____________.

A. Documentation of errors, workarounds, and test information

B. Training and knowledge transfer

C. Visualizing knowledge management

D. Protection of intellectual property

36. Which of the following is not an availability measurement perspective?

A. IT Service Provider

B. User

C. Technical

D. Business

37. Which item best reflects a Critical Success Factor in Service Catalog Management?

A. Users are familiar with the services delivered

B. Accurate Service Catalog

C. IT Organization is familiar with the techniques which support the service.

D. All of the above

38. Which of the following would NOT be an attribute of a CI?

A. Installation instructions

B. Purchase price

C. Comments

D. Value after depreciation

39. Which of the following define how change will be managed, including roles, responsibilities, and escalation processes?

A. Change log

B. Change control

C. Configuration management

D. Change model

40. What questions should an organization answer while developing a service strategy?

A. What products and services will we offer?

B. What capabilities and capacities will we require?

C. What is the current strategy?

D. All of the above

ITIL Foundation Practice Exam 6 Answer Key and Explanations

1. C - Use the four Ps of Personnel, Processes, Products (including technology), and Partners (including suppliers) as a standard building block to plan and provide a service. [ITIL - Technology and Architecture]

2. C - The reliability of a service is reported as the mean time between failures (MTBF) or uptime. [ITIL Generic Concepts and Definitions]

3. D - The ability of a third-party supplier to meet the terms and conditions of their contract is known as serviceability. Serviceability is based on how well the supplier meets agreed upon levels of reliability, maintainability, and availability. [ITIL - Generic Concepts and Definitions]

4. C - The five aspects of Service Design are design of Service Solution, Systems, Technology Architecture, Processes, and Measurement Systems and metrics. [ITIL - Selected Processes]

5. B - Ongoing alignment of services with current and future business needs is one of the objectives of the Continual Service Improvement phase. [The Service Lifecycle]

6. B - If process structures become an objective in themselves, the service quality may be adversely affected; unnecessary or over-engineered procedures are seen as bureaucratic obstacles, which are to be avoided where possible. [ITIL - Service Lifecycle]

7. B - The Mean time to Restore Service (MTRF) is the time within which a function is back up after a failure. [ITIL - Selected Functions]

8. B - Tension metrics are a set of related metrics, in which improvements to one metric have a negative effect on another. Tension metrics are designed to ensure an appropriate balance between conflicting objectives. [ITIL - Generic Concepts and Definitions]

9. C - A configuration management system (CMS) is the collection of all tools, databases, and information management systems used to store and manage configuration items (CIs). A configuration management database (CMDB) is part of the configuration management system. [ITIL - Key Principles and Models]

10. B - A successful implementation requires the involvement and

commitment of personnel at all levels in the organization; leaving the development of the process structures to a specialist department may isolate that department in the organization and it may set a direction that is not accepted by other departments. [ITIL - Service Lifecycle]

11. D - Supplier metrics are defined in the Supplier Management process, not in Continual Service Improvement (CSI) [ITIL - Selected Processes]

12. B - One of the key roles of Service Operation is achieving balance between procedures and activities taking place in a continually changing environment. [ITIL - Selected Roles]

13. D - The Service Design phase in the lifecycle begins with the demand for new or modified requirements from the customer. [ITIL - Selected Processes]

14. B - The dominant pattern in the Service Lifecycle is the succession of Service Strategy to Service Design, to Service Transition and to Service Operation, and then, through Continual Service Improvement, back to Service Strategy, and so on. [ITIL - Key Principles and Models]

15. B - Service Strategy is the axis of all other phases of the Service Lifecycle. [ITIL - Service Lifecycle]

16. D - The output from every phase is input for the next phase of the Service Lifecycle. Thus, Service Strategy is the input to Service Design. [ITIL - Service Lifecycle]

17. B - The four phases of the Deming Cycle are Plan, Do, Check, and Act. The Deming Cycle is also known as the PDCA Cycle. [ITIL Key Principles and Models]

18. B - The OLA supports the service providers' delivery of services to customers by defining the goods or services to be provided and the responsibilities of both parties. [ITIL - Key Principles and Models]

19. B - Service design is constrained by internal resources such as available staff or funding, and by external drivers such as legislative or regulatory controls the organization is subject to. [ITIL - Selected Processes]

20. D - Every requirement must be SMART (Specific, Measurable, Achievable/ Appropriate, Realistic/ Relevant, and Timely/ time-bound) formulated. [ITIL - Selected Processes]

21. C - Service requests are initiated by a single user requiring assistance with some aspect of an IT service, such as replacement of a failed hard drive. New or enhanced services provided simultaneously to many customers, such as increased email attachment sizes or application functionality, are driven by Service Strategy. [ITIL - Key Principles and Models]

22. D - A baseline is a snapshot used to create a starting point for measuring performance or as a foundation for further development. Infrastructure configurations and performance metrics are both candidates for baseline creation. [ITIL - Key Principles and Models]

23. A - Service value is the sum of Service Utility (fitness of purpose, or what the service does) and Service Warranty (fitness of use, or how well the service performs). [ITIL Key Principles and Models]

24. D - Business relationship management is the process by which customer needs are identified. This process is also responsible for ensuring that these needs are met. [ITIL Key Principles and Models]

25. A - The Service Design phase processes are service catalog management, service level management, capacity management, availability management, IT service continuity management, information security management, supplier management. [ITIL - Selected Processes]

26. C - A SLAM chart is typically color coded to show whether each agreed Service Level Target has been met, missed, or nearly missed during each of the previous 12 months. [ITIL - Selected Functions]

27. C - The Service manager co-ordinates the development, introduction, and evaluation of one or more products or services. He or she is responsible for achieving company strategy and goals, benchmarking, financial management, customer management, and vendor management. [ITIL - Selected Roles]

28. B - A procedure is a specified way to carry out an activity or a process. It describes the "how" and can also describe "who" carries out the activity. A procedure may include stages from different processes. A "work instruction" also describes the steps needed to carry out an activity, but at a much greater level of detail. [ITIL - Generic Concepts and Definitions]

29. C - A standard change is a change of a service or infrastructure component that change management must register, but is of low risk and is pre-authorized. These are routine changes, such as PC upgrades. [ITIL - Selected Functions]

30. C - The change manager is responsible for receiving, logging, and prioritizing requests for change. [ITIL - Selected Roles]

31. D - Bundling core services and supporting services are a vital aspect of a market strategy. Service providers should thoroughly analyze the primary conditions in their business environment, the needs of the customer segments or types they serve, and the alternatives available to these customers. [ITIL - Selected Functions]

32. A - The likelihood that the risk will occur and its possible impact determine the risk category of the change. In practice, a risk categorization matrix is generally used for this purpose. [ITIL - Selected Functions]

33. D - Wisdom comes from knowledge, experience, and judgment. It allows organizations to add value to its services by understanding why specific events happen. [ITIL Key Principles and Models]

34. A - The V model is a convenient tool for mapping out the different configuration levels at which building and testing must take place. The left side of the V starts with service specifications and ends with the detailed Service Design. The right hand side of the V reflects the test activities, by means of which the specifications on the left-hand side must be validated. In the middle we find the test and validation criteria. [ITIL - Key Principles and Models]

35. C - Knowledge management is often visualized using the DIKW structure: Data-Information-Knowledge-Wisdom. [ITIL - Key Principles and Models]

36. C - Measuring availability can be done from three perspectives: business, user, and the IT service provider. [ITIL - Selected Functions]

37. D - Critical Success Factors for Service Catalog Management are: accurate Service Catalog, users are familiar with the services delivered, and the IT Organization is familiar with the techniques which support the service. [ITIL - Selected Functions]

38. A - Attributes are identifying information about a CI. Installation instructions would not be an attribute, but rather would be controlled as a CI. [ITIL - Technology and Architecture]

39. D - A change model defines how changes are to be handled. Each organization must define its own change model; there may be multiple change models in place in a single organization to support a variety of change types. [ITIL - Key Principles and Models]

40. D - The goal of Service Strategy is to identify the competition and to compete with them by distinguishing oneself from the rest and by delivering superior performances. [ITIL - Key Principles and Models]

ITIL Foundation
Practice Exam 7
Practice Questions

Test Name: ITIL Foundation Practice Exam 7
Total Questions: 40
Correct Answers Needed to Pass: 30 (75.00%)
Time Allowed: 60 Minutes

Test Description

This is the seventh cumulative ITIL Foundation test which can be used as an indicator for overall performance. This practice test includes questions from key ITIL areas.

Test Questions

1. The Smith Company manufactures office furniture. They have contracted with Jones Consulting to handle all server backups and restores. This is an example of what service management principle?

    A. Agency principle

    B. Systems

    C. Encapsulation

    D. Specialization and co-ordination

2. Which of the following is not a supplier category as defined by ITIL

    A. Internal

    B. Tactical

    C. Operational

    D. Strategic

3. A policy requiring that all improvement initiatives must go through the change management process is an example of what type of policy?

    A. Service Design

    B. Regulatory

    C. Continuous Service Improvement

    D. Advisory

4. ________________is a repeatable model of dealing with a particular Category of Change.

    A. Change Control

    B. Change Model

C. Change Management

D. A Standard Change

5. The main continuous service improvement activities are

A. Review, discuss, improve

B. Report, design, improve.

C. Check, report, improve.

D. Check, report, implement

6. Fill in the blank: ____________________ is where the approved versions of all media CIs are stored and monitored.

A. CMDB

B. Definitive Media Library

C. Definitive spares

D. Secure library

7. Availability management monitors, measures, analyzes, and reports on:

A. Failure rates, recovery method, policy compliance, and process compliance.

B. Availability, redundancy, manageability, and service effectiveness

C. Availability, reliability, maintainability, and serviceability

D. Component function, output, utilization, and mean time between failures.

8. Information Security Management is concerned with the maintenance of what elements of an organization's information assets?

A. Confidentiality, integrity, availability

B. Privacy, redundancy, criticality

C. Classification, access, confidentiality

D. Encryption, loss prevention, risk management

9. No change should be approved without:

A. Consulting senior management

B. A fallback plan

C. Knowing who can legally authorize the change.

D. Reviewing the corporate disaster recovery plan

10. What function keeps track of all the interfaces and dependencies between all present services and those under development?

A. Configuration Management

B. Continual Service Improvement

C. Service Catalogue Management

D. Service Management

11. Fill in the blank: ______________ is the addition, modification, or elimination of an authorized, planned, or supporting service and its related documentation.

A. Baseline review

B. Continual Service Improvement

C. Change

D. Service asset and configuration management

12. A pilot launch occurs during which phase?

A. Continual Service Improvement

B. Service Design

C. Service Transition

D. Service Operation

13. Superbrands Stores is rolling out a new pay-per-use Help Desk to supplement the existing web-based self-help system. During the pilot phase, Superbrands departments are alerted to the total charges incurred by their users each week, but no money is actually transferred to cover the cost of the service. This is an example of :

A. Control Loop Charging

B. Notional Charging

C. Internal Chargeback's

D. Virtual Charging

14. What is the service support interface point between a service and its users?

A. The service desk

B. Customer service department

C. A web page

D. none of the above

15. Who is ultimately responsible for ensuring a service meets a client's requirements?

A. Process Owner

B. Upper management

C. Service Owner

D. Customer Service

16. A large multinational corporation has outsourced several IT functions to different service providers. Each service provider maintains its own CMS, which is subsequently shared with the customer. This is an example of what?

A. Multi-sourcing

B. Organizational strategy

C. Data integration

D. Federated CMDB

17. An especially dangerous worm has been identified. It has been determined that a security patch must be installed on all vulnerable systems as soon as possible, outside the normal service window for these devices and/or services they support. What is this type of update called?

A. Emergency change

B. Proactive change

C. Reactive change

D. Risk management

18. Once an IT organization learns the overall organization's vision, it will form a strategy that aligns with the business strategy. What is the next step?

A. Define the service improvement plan

B. Define targets

C. Develop policies

D. Document the current situation.

19. The Service Portfolio represents all the resources that are active in the various phases of the ____________________.

A. Planning Process

B. Service Lifecycle

C. Service Strategy

D. Development Lifecycle

20. IT Service Continuity Management would not focus on:

A. An unplanned shutoff of water service to the call center due to storm damage in the area.

B. Installation of cubicles in a disaster recovery data center.

C. Merging the IT departments of two different organizations as part of a corporate acquisition.

D. Email system failure

21. An Incident occurs when:

A. A portion of a network has failed, however it is not noticeable to users because of built in redundancy.

B. An employee calls the help desk to report the system for entering work hours is very slow

C. A Customer Service Rep can't access a needed application

D. All of these items are correct

22. The main challenge in implementing IT Security Management is:

A. Strict change management and configuration management.

B. Obtaining adequate support from the company, business security, and senior management.

C. Establishing clear policies

D. Justifying the need for IT Security Management

23. What term indicates "fitness for use" ?

A. Liability

B. Guarantee

C. Warranty

D. Utility

24. What type of asset is controlled by SACM?

A. Service Level Agreement

B. Software only

C. Hardware only

D. Configuration Item

25. Worldwide Mergers and Acquisitions is rolling out a new service offering. In which phase will analysis of the service offering be most useful?

A. Measurement phase.

B. Planning phase

C. Implementation phase

D. All of these responses / All of the above

26. What is the most important challenge in Service Catalog Management?

A. Maintaining the Service Knowledge Management System.

B. Maintaining an accurate Service Catalog.

C. Implementing a Configuration Management System.

D. Achieving KPIs

27. A process is a logically coherent series of activities for a pre-defined goal. What is the process owner responsible for?

A. Describing the process

B. Setting up the process

C. The result of the process

D. Implementing the process

28. Status, Root Cause, and Workaround are the main phases of ____________________

A. Event management

B. Incident management

C. Root Cause analysis

D. Lifecycle of a Known Error

29. Supporting business continuity by ensuring that required IT facilities can be restored within an agreed time is the focus of:

A. Disaster recovery planning

B. IT service continuity management

C. Backup and restore processes

D. Incident management

30. An easily quantifiable impact caused by loss of service, such as a loss of revenue, is called a hard impact. A soft impact is less easily quantifiable. Which of the following is an example of a soft impact?

A. Increase in facility rent

B. Damage to corporate reputation

C. Shortage of hot spares

D. Increase in the cost of delivering services

31. Gary is responsible for managing the media for his organization's data backup system. The activities that he performs include changing media in the robotic tape library, ordering new media as required, and ensuring that backup media is archived according to the organization's policies. What is Gary's role?

A. Process practitioner

B. Process owner

C. Operational practitioner

D. Operational owner

32. Business impact analysis involves identifying the critical business functions within the organization and determining the impact of failure to perform the business function beyond the maximum acceptable outage. What types of criteria can be used to evaluate this impact?

A. Policy and process

B. Exposure and liability

C. Internal and external risks

D. Customer service and finance

33. Measurement and analysis of SLA achievement is a main objective of:

A. IT Service management processes

B. Continuous Service Improvement

C. Quality control

D. The service owner

34. Fill in the blanks: ____________________ represents all active and inactive services, while ____________________ consists only of services available at the retail level.

A. Service Catalog, Service Portfolio

B. Line of Service, Service Package

C. Service Portfolio, Service Catalog

D. Service Package, Line of Service

35. Adding additional support capacity, such as extended help desk hours, often increases the demand for additional support mechanisms such as a self-service webpage. The increased use of the support service justifies the costs associated with maintenance and upgrades, further enhancing the potential for better performance. This is an example of:

A. Management by design

B. Good planning

C. A closed-loop control system

D. An adaptive system

36. Which of the following roles is responsible for recommending service improvements?

A. Process owner

B. Product manager

C. CSI manager

D. Service owner

37. The ________________ is the maximum amount of data that may be lost when Service is Restored after an interruption.

A. Recovery Point Objective

B. Mean Time Between Failures

C. Service Level Target

D. Recovery Time Objective

38. Which capacity management sub process focuses on the performance and utilization of disk, CPU, and memory utilization in a server?

A. Service operation agreement

B. Onsite Service Agreement

C. Component capacity management

D. Service capacity management

39. There are several Key Performance Indicators in the Service Design process. Which of the following is not a Service Design process KPI?

A. Accuracy of the Service Catalog

B. Accuracy of the SLA.

C. Percentage of specifications of requirements of Service Design produced within budget.

D. Percentage of specifications of requirements of Service Design produced on time.

40. What is a RACI model used for?

A. Recording configuration items

B. Monitoring services

C. Performance analysis

D. Define roles and responsibilities

ITIL Foundation Practice Exam 7 Answer Key and Explanations

1. D - The goal of service management is to make capabilities and resources available through services that are useful and acceptable to the customer with regard to quality, costs, and risks. The service provider takes the weight of responsibility and resource management off the customer's shoulders so that they can focus on the business' core competence. [ITIL - Generic Concepts and Definitions]

2. A - Internal is not a valid supplier category. ITIL categorizes suppliers as strategic, tactical, operational, or commodity. [ITIL Key Principles and Models]

3. C - CSI Policies capture agreements about measuring, reporting, service levels, CSFs, KPIs, and evaluations. [ITIL - Selected Processes]

4. B - A Change Model includes specific pre-defined steps that will be followed for a Change of this Category. [ITIL - Generic Concepts and Definitions]

5. C - To improve the services of the IT organization, CSI measures the yield of these services. The main CSI activities are check, report, and improve. [ITIL - Selected Processes]

6. B - The Definitive Media Library is a secure store where the definitive, authorized versions of all media CIs are stored and monitored. [ITIL - Technology and Architecture]

7. C - Availability management monitors, measures, analyzes, and reports on availability, reliability, maintainability, and serviceability. [ITIL - Selected Functions]

8. A - Information Security Management is concerned with protecting the confidentiality, integrity, and availability of an organization's information assets. [ITIL Selected Functions]

9. B - No change should be approved without having an answer to the following question:"what will we do if the change is unsuccessful?" You must always ensure that a fallback situation (remediation plan) is available. [ITIL - Selected Functions]

10. C - The purpose of Service Catalogue Management (SCM) is the development and upkeep of a Service Catalogue that contains all accurate details, the status, possible interactions and mutual dependencies of all present services and those under

development. [ITIL - Selected Functions]

11. C - Change is the addition, modification, or elimination of an authorized, planned, or supporting service and its related documentation. [ITIL - Selected Functions]

12. C - Service Transition includes the management and coordination of the processes, systems, and functions required for the building, testing, and deployment of a release into production. [ITIL - Service Lifecycle]

13. B - Notional charging is an approach to charging for IT services, where charges are calculated and customers are informed of the charge, but no money is actually transferred. [ITIL - Key Principles and Models]

14. A - The service desk is the single interface point between a service and its users. Customer service departments and web pages are tools used by the service desk to provide assistance to the user base. [ITIL Selected Functions]

15. C - The service owner ensures that the service meets the requirements. [ITIL - Selected Roles]

16. D - At the data level it may be that the CMS gets its data from different CMDBs, that together, form a federated CMDB. [ITIL - Selected Functions]

17. A - An emergency change is one that takes place as soon as possible, and outside the normal change window. Emergency changes are not always critical updates such as security patches. They may also be changes requested to support a project. For example, a development group may need to implement a newly identified firewall rule in a staging environment to permit application testing, and the application schedule will be unacceptably delayed if the team must wait till the next regular change window for the rules to be implemented. [ITIL - Selected Processes]

18. D - The CSI model consists of 6 phases: determine the vision, record the current situation, determine measurable targets, plan, check, and assure. [ITIL - Selected Processes]

19. B - The service portfolio represents the opportunities and readiness of a service provider to serve the customers and the market. [ITIL - Service Lifecycle]

20. C - ITCSM focuses on those events that can be considered a disaster, not small technical problems that are handled by incident management. [ITIL - Selected Functions]

21. D - An incident is any unplanned interruption to or reduction in quality of an IT service, therefore all are correct. [ITIL - Generic Concepts and Definitions]

22. B - The main challenge in the implementation of IT Security Management is to ensure adequate support of the company, business security, and senior management. If this is missing, it is impossible to establish an effective security process. [ITIL - Selected Functions]

23. C - ITIL uses two important concepts for the value of a service. For customers, the positive effect is the "utility" of a service; the insurance of the positive effect is the "warranty". [ITIL - Generic Concepts and Definitions]

24. D - A configuration item is an asset, service component, or other item that is (or will be) controlled by configuration management. [ITIL - Selected Functions]

25. D - Assessments are useful in the planning, implementation, and measurement phases. [ITIL - Selected Functions]

26. B - The most important challenge in the Service Catalog Management process is maintaining an accurate Service Catalog (containing both the Business and Technical Aspect) as part of the Service Portfolio. [ITIL - Selected Functions]

27. C - The process owner is responsible for ensuring that the process is implemented as agreed and that the established objectives will therefore be achieved. [ITIL - Selected Roles]

28. D - A Known Error Record documents the Lifecycle of a Known Error, including the Status, Root Cause, and Workaround. [ITIL - Selected Functions]

29. B - The ultimate goal of IT service continuity management is to support business continuity by ensuring that the required IT facilities can be restored within the agreed time. [ITIL - Selected Processes]

30. B - Business impact analyses quantify the impact caused by the loss of services. If the impact can be determined in detail, it is called a hard impact. If it is less easily determined, it is called a soft impact. [ITIL - Selected Processes]

31. A - The process practitioner is responsible for performing one or more process activities. This role is also responsible for ensuring inputs and outputs for the processes are correct, and creating/ managing activity records. [ITIL Selected Roles]

32. D - The Business Impact analysis is an essential element in the business continuity process and dictates the strategy to be followed for risk reduction and recovery after a catastrophe. The BIA consists of two parts: investigation of the loss of a process or function, and eliminating the effect of that loss. [ITIL - Selected Processes]

33. B - A main objective of CSI is to measure and analyze Service Level Achievements by comparing them to the requirements in the SLA. [ITIL - Selected Functions]

34. C - The Service Portfolio represents all active and inactive services; the Service Catalog contains only active and approved services at retail level. [ITIL - Selected Functions]

35. C - Service management is a closed-loop control system with the following functions: developing and understanding service assets, understanding the performance potential of customer assets, mapping of service assets to customer assets through services, and designing, developing, and adapting services. [ITIL - Generic Concepts and Definitions]

36. D - The service owner is responsible for the continual improvement of the services he or she has been assigned. This includes the recommendation of improvement or changes to services as needed. [ITIL Selected Roles]

37. A - The Recovery Point Objective is the maximum amount of data that may be lost when Service is Restored after an interruption, and is a key concept within Business Continuity Planning. [ITIL - Generic Concepts and Definitions]

38. C - Component capacity management is the sub process focusing primarily on managing, controlling and predicting performance, with the emphasis on IT infrastructure that supports a service. [ITIL - Selected Processes]

39. A - The Service Catalog is a prerequisite of the SLA. Before Service Level Management can design the SLA, a business service catalog and technical service catalog are necessary. [ITIL - Key Principles and Models]

40. D - The RACI matrix is used to define responsibility and accountability.

RACI stands for responsible, accountable, consulted, informed. [ITIL - Key Principles and Models]

ITIL Foundation
Practice Exam 8
Practice Questions

Test Name: ITIL Foundation Practice Exam 8
Total Questions: 40
Correct Answers Needed to Pass: 30 (75.00%)
Time Allowed: 60 Minutes

Test Description

This is the eighth cumulative ITIL Foundation test which can be used as an indicator for overall performance. This practice test includes questions from key ITIL areas.

Test Questions

1. A tax preparation firm deployed a new application for its accountants to use. While it meets the needs of the firm's staff, it has a number of security vulnerabilities that were subsequently exploited, potentially exposing personally identifiable information to unauthorized users. What type of requirement was overlooked during the service design process?

   A. Management/ Operational

   B. Administrative

   C. Usability

   D. Functional

2. When must Service Level Requirements be defined?

   A. As part of the contract negotiation process.

   B. Before the service is delivered.

   C. Every time the service is performed.

   D. After the service is delivered.

3. A law firm has recently implemented an automated tool to allow users to reset forgotten or expired passwords. This is an example of what?

   A. Automation

   B. Agency Principle

   C. Encapsulation

   D. Single Sign On

4. Which opportunity analysis tool uses a 4-part table to compare internal and

external conditions that could affect the implementation?

A. SWOT

B. Rummler-Brache Swim Lane Diagram

C. Waterfall Model

D. Balanced Score Card

5. Which of the following is a technical management metric?

A. Budget variance

B. Pending RFCs

C. Mean time between failures (MTBF)

D. Average time to ticket resolution

6. Which approach to Pre-Program ROI is the best choice?

A. Internal Rate of Return

B. Net Present Value

C. Accrual Accounting

D. Cashflow Basis

7. A Configuration Management Database would be used during with ITSM phase?

A. Catalog Definition

B. Service Delivery

C. Service Transition

D. System Transition

8. What process is being conducted by documenting and subsequently updating the configuration of the backup server and tape library that performs backups for a shared IT service?

A. Service Asset and Configuration Management

B. Service Offering Definition

C. Change Control

D. Service Catalog Management

9. Planning for, conducting, and providing guidance for a service deployment and the release of all documentation is the responsibility of which role?

A. Configuration Manager

B. Deployment Manager

C. Change Advisory Board

D. Change Manager

10. ABC Plastics has decided to move its IT services from a 3rd party vendor to a shared services group within it's parent company. What is this type of change called?

A. Aggregation

B. Insourcing

C. Functional reorganization

D. Corporate reorganization

11. Which of the following activities is not part of the Service Strategy process?

A. Preparation for implementation

B. Development of the offer

C. Development of strategic assets

D. Value net configuration

12. Shock, avoidance, blame, self-blame, and acceptance are phases of what event?

A. Emotional Change Cycle

B. Damage Control

C. Service Failure

D. Emergency Rollback

13. Michael is responsible for specific services at his company. He attends Change Advisory Board meetings when they are relevant to his services, he maintains the service description in the Service Catalog, and he measures the performance and availability of the service. What is Michael's role called?

A. IT Manager

B. Service Catalog Manager

C. Service Owner

D. Process Owner

14. Which implementation approach usually results in an unsuccessful program?

A. The 4 P's

B. Timeslicing

C. Kotter Method

D. Big Bang

15. Designing a service based on customer preferences, using an approach such as the Kano Model is called ________________.

A. Outcome-oriented design

B. Asset-based design

C. Market-based design

D. Constraint- oriented design

16. Sometimes referred to as black-boxing, this service delivery principlre shields customers from the backend details of the service being delivered.

A. Rabbit in the hat

B. Condensed view

C. Encapsulation

D. Oblique delivery

17. A team of DBAs is troubleshooting an outage on the main finance database. The outage was reported by a branch office that was unable to access the latest billing information. The investigation of the underlying cause of the outage is known by what term?

A. Incident Management

B. Problem Management

C. Incident Reporting

D. Outage Management

18. DataCorp provides webhosting services to all its subsidiaries through its internal IT division . This is an example of what type of Service Provider?

A. Type II

B. Outsourcing

C. Expansion-based

D. Managed Hosting

19. Jones, Inc, a large mail order company, is determining what sort of improvements it needs to make to its CRM platform to better support its sales staff and its customers. How will raw technical, process, performance, and value metrics be converted into supporting evidence to help decision makers with their planning?

A. RACI

B. PCDA

C. DIKW

D. MTBF

20. ITIL defines three different types of information that determine the objectives of a service. What are they?

A. Tasks, Results, Constraints

B. Plans, Warranty, Profit

C. Action, Reaction, Constraints

D. Goal, Organization, Value

21. Authentication, authorization, accounting, and auditing are tasks within what process?

A. Access Management

B. Identity Protection

C. Policy Definition

D. Security Management

22. The which of the following is not a CSI metric?

A. Logging Metrics

B. Process Metrics

C. Technology Metrics

D. KPIs

23. After XYZ Corp rolled out its new Help Desk offering, it analyzed the KPIs of the new offering and compared them to the metrics of the old offering to determine how successful the new Help Desk is. What is this called?

A. Isolation of the program's effects

B. Program cost analysis

C. Value based analysis

D. Post-service planning

24. Which of the following is monitored as part of Continual Service Improvement (CSI)?

A. Process compliance

B. Quality

C. Performance

D. All of the above

25. What does the acronym RACI stand for?

A. Responsible, Accountable, Consulted, Informed

B. Requirements, Availability, Computers, Information

C. References, Acceptance, Control, Integration

D. Risks, Actions, Completion, Investigation

26. Connecting Point provides an outsourced help desk and call center services for a variety of corporate clients. A report is created at the end of each shift, and the findings are presented at a shift change meeting, so the incoming staff is aware of issues or incidents that arose during the previous shift. This is an example of what type of Service Operation function?

A. Change Management

B. Service Strategy

C. Communication

D. Service Advisory Bulletin

27. The actual deployment of a new or updated service is actively managed by which role?

A. Service Catalog Manager

B. Service Transition Manager

C. Configuration Manager

D. Service Owner

28. Please select the most important Continual Service Improvement process:

A. Post Implementation Review

B. Service Catalog Update

C. Contract Negotiation

D. Service Level Management

29. In which phase would the Denali Integration Partners' Service Strategy team ask the question "have we reached our intended goals?"

A. Service Definition

B. Service Strategy

C. Continual Service Improvement

D. Service Support

30. Balancing business requirements with services delivery requirements and constraints is a function of what lifecycle phase?

A. Service Design

B. Service Strategy

C. Service Delivery

D. All phases

31. Jennifer is an engineer director at a managed hosting company. She is in charge of the processes by which new patches will be deployed, and works with other engineers to design the most effective mechanisms for the patching efforts. In what role is Jennifer acting?

A. Service Owner

B. Process Owner

C. Management representative

D. Service Deployment

32. By what element or process is the Service Portfolio managed?

A. Service Design

B. Portfolio Management

C. Service Strategy

D. Service Catalog Management

33. The executive leadership of Acme Widgets has decided to implement a new CRM platform to provide better service to existing customers and help sales teams identify opportunities based on trending reports. In which phase of the ITIL Lifecycle will these goals be defined?

A. Continual Service Improvement

B. Service Design

C. Service Operation

D. Service Strategy

34. Deming's PCDA cycle is used to manage what phase of the ITIL lifecycle?

A. Service Operation

B. Continuous Service Improvement

C. Continual Service Improvement

D. Service Implementation

35. Using a tool such as TIBCO Hawk to monitor the behavior of enterprise applications and manage these applications to keep services available at all times is an example of what Service Operation function?

A. Monitoring and Control

B. Service Management

C. Service Support

D. Plan, Do, Check, Act

36. The ongoing maintenance and support of technical components such as mainframes, desktop computers, and databases falls under which lifecycle phase?

A. Service Operation

B. Service Support

C. Monitoring and Control

D. Service Management

37. What is a disadvantage of a Type 1 Service Provider?

A. Short lines of communication

B. Specialized in a limited set of business needs

C. Limited decision rights

D. Growth is tied to the growth of the business unit

38. Which role is responsible for ensuring that the Service Catalog is in sync with the Service Portfolio?

A. The Service Design Manager

B. The Availability Manager

C. The Service Catalog Manager

D. The Service Level Manager

39. A global semiconductor company wants to implement a shared services IT group for the entire organization.

The have listed their goals and would like to determine the underlying processes associated with them so they can understand all the changes that would be required for a successful implementation. What strategic model would be the best for this purpose?

A. Gap Analysis

B. Iterative Model

C. Waterfall Model

D. Balanced Score Card

40. Which process tool shows the flow of a process through an organization and the interface between people, departments, and technology?

A. Waterfall Model

B. Gap Analysis

C. SWOT

D. Rummler-Brache Swim Lane Diagram

ITIL Foundation Practice Exam 8 Answer Key and Explanations

1. A - Organizations often expend a great deal of time on the functional requirements of a new service, while very little time is spent defining the management and operation requirements of the service. The result is a service that fails to meet expectations in the areas of quality and performance. ITIL - Selected Processes [ITIL - Selected Processes]

2. B - In order to ensure services meet customer expectations, Service Level Requirements must be clear before the service is delivered. ITIL - Selected Processes [ITIL - Selected Processes]

3. B - Service Management always involves an agent and a principal that seconds this agent to fulfill activities on their behalf. Agents may be consultants, advisors, or service providers. Usually these agents are the service provider's staff, but they can also be self-service systems and processes for users. ITIL - Service Management as a Practice [ITIL - Service Management as a Practice]

4. A - SWOT identifies strength and weakness (internal conditions), as well as opportunity and threats (external conditions) that may impact an organization's ability to meet its goals. ITIL - Key Principles and Models [ITIL - Key Principles and Models]

5. C - Mean time between failure is a measurement of the interval between failures of a given system component, usually hardware. ITIL - Technology and Architecture [ITIL - Technology and Architecture]

6. B - Net Present Value is based on the comparison between cash inflows and cash outflows, where the difference, the "net present value", determines the whether or not the investment is valuable. Net Present Value makes a more realistic assumption for the rate of return. ITIL - Key Principles and Models [ITIL - Key Principles and Models]

7. C - Technology plays an important part of the support of Service Transition, and includes systems such as the Configuration Management Database, network management and monitoring tools, and automated deployment systems. ITIL - Technology and Architecture [ITIL - Technology and Architecture]

8. A - The objective of service asset and configuration management is the definition of service and infrastructure

components and the maintenance of accurate configuration records. ITIL - Selected Processes [ITIL - Selected Processes]

9. B - The Deployment Manager is responsible for the final service implementation, from delegation to wrap-up metric reporting. ITIL - Selected Roles [ITIL - Selected Roles]

10. B - Migration from a Type III to a Type II service provider is called insourcing. InformIT- ITIL - Generic Concepts and Definitions [ITIL - Generic Concepts and Definitions]

11. D - The four most important activities of the Service Strategy Process are defining the market, development of the offer, development of strategic assets, and preparation for implementation. ITIL - Selected Processes [ITIL - Selected Processes]

12. A - One of the biggest causes of failure of changes is not sufficiently considering the way in which the changes affect people. The emotional phases that may occur before change acceptance are shock, avoidance, external blame, self blame, and acceptance. ITIL - Generic Concepts and Definitions [ITIL - Generic Concepts and Definitions]

13. C - The Service Owner is the primary point of contact for a specific service, and owns and represents that service for the organization. ITIL - Selected Roles [ITIL - Selected Roles]

14. D - A big bang approach does not usually result in a successful improvement program. Step by step approaches such as Deming's PCDA are more successful. ITIL - Key Principles and Models [ITIL - Key Principles and Models]

15. A - Outcome oriented design develops a service from the customer's perspective and preferences. ITIL - Key Principles and Models [ITIL - Key Principles and Models]

16. C - The encapsulation principle says a product or service offering should hide what the customer does not need to know about technical details while showing what is valuable and useful to the customer. ITIL - Service Management as a Practice [ITIL - Service Management as a Practice]

17. B - Problem management involves analyzing and resolving the causes of incidents. ITIL - Generic Concepts and Definitions [ITIL - Generic Concepts and Definitions]

18. A - The market of Type II Service Providers is internal to the enterprise, but it is distributed through the business units. ITIL - Generic Concepts and Definitions [ITIL - Generic Concepts and Definitions]

19. C - DIKW is an acronym for Data, Information, Knowledge, and Wisdom. Metrics supply quantitative data. CSI transforms this data into qualitative information. Combining Information with experience, context, and interpretation, it becomes knowledge. Using DIKW, the leadership team can determine what enhancements should be implemented to meet the needs of internal and external customers. ITIL - Selected Processes [ITIL - Selected Processes]

20. A - The three types of information that determine the objectives of a service are tasks, results, and constraints. ITIL - Selected Processes [ITIL - Selected Processes]

21. A - Access management is the process of allowing permitted users to access services, while preventing access by users without appropriate permission. ITIL - Selected Processes [ITIL - Selected Processes]

22. A - Three types of metrics are needed for CSI: technology metrics, process metrics, and KPIs (key performance indicators). ITIL - Generic Concepts and Definitions [ITIL - Generic Concepts and Definitions]

23. A - Isolation of the program's effects uses various techniques to ensure the program is effective, including determining what would have occurred if the program had not been started. ITIL - Selected Processes [ITIL - Selected Processes]

24. D - CSI monitors and measures process the compliance, quality, performance, and business value of a process. ITIL - Selected Processes [ITIL - Selected Processes]

25. A - RACI is an acronym for the four most important roles to be defined in a service. Each letter stands for an answer to the question Who is Responsible, Accountable, Consulted, and Informed. ITIL - Selected Roles [ITIL - Selected Roles]

26. C - Good communication can prevent problems, and every team and department must have a clear communications policy. Service Operation has a variety of communication types, including routine operational communication and communication between shifts. ITIL - Selected Functions [ITIL - Selected Functions]

27. B - Service Transition is actively managed by a Service Transition Manager, who is responsible for the daily management and control of the Service Transition teams and their activities. ITIL - Selected Roles [ITIL - Selected Roles]

28. D - Service Level Management (SLM) is the most important process for CSI. SLM helps the business and the IT organization understand what needs to be measured and what the results should be. ITIL - Selected Processes [ITIL - Selected Processes]

29. C - As part of Continual Service Improvement, an organization may conduct an implementation review to determine whether improvements have produced the desired effects. ITIL - Service Lifecycle [ITIL - Service Lifecycle]

30. D - The external business view versus the internal IT view is usually the key conflict in all phases of the ITSM lifecycle. The issues are typically stability versus responsiveness, reactive versus proactive, and cost versus quality. ITIL - Service Lifecycle [ITIL - Service Lifecycle]

31. B - The process owner is responsible for the process strategy, assists in the design, and ensures that all process activities are carried out. ITIL - Selected Roles [ITIL - Selected Roles]

32. C - Although the Service Portfolio is designed during the Service Design phase, the portfolio is managed by the Service Strategy. ITIL - Service Lifecycle [ITIL - Service Lifecycle]

33. D - A Service Strategy helps in identifying, selecting, and prioritizing opportunities. ITIL - Service Lifecycle [ITIL - Service Lifecycle]

34. C - The Plan- Do-Check-Act cycle prescribes a repeating pattern of improvement efforts. ITIL - Service Lifecycle [ITIL - Service Lifecycle]

35. A - Service monitoring and control is based on a continual cycle of monitoring, reporting, and undertaking action. This cycle is crucial to providing, supporting, and improving services. ITIL - Selected Functions [ITIL - Selected Functions]

36. A - Technical management of the infrastructure and components behind the services being offered is part of Service Operations. ITIL - Service Lifecycle [ITIL - Service Lifecycle]

37. D - The disadvantage of a Type 1 Service Provider is limited opportunities for growth, as growth is tied to the growth of the business unit.

ITIL - Generic Concepts and Definitions [ITIL - Generic Concepts and Definitions]

38. C - The Service Catalog manager is responsible for the production and maintenance of the Service Catalog, as well as ensuring that information is consistent with the information in the Service Portfolio. ITIL - Selected Roles [ITIL - Selected Roles]

39. D - The Balanced Scorecard is a performance management tool for measuring whether the smaller scale operational activities of a company are aligned with its larger scale objectives in terms of vision and strategy. ITIL - Key Principles and Models [ITIL - Key Principles and Models]

40. D - A Rummler-Brache Swim Lane Diagram maps the flow of a process. Horizontal rows divide the separate organizations or departments from each other. Activities and decisions are connected through arrows to indicate flow. ITIL - Key Principles and Models [ITIL - Key Principles and Models]

ITIL Practice Test: Fill in the Blank Practice Questions

Test Name: ITIL Practice Test: Fill in the Blank
Total Questions: 20
Correct Answers Needed to Pass: 15 (75.00%)
Time Allowed: 15 Minutes

Test Description

This is a fill-in-the-blank practice test based on multiple ITIL knowledge areas. This practice test focuses on understanding ITIL terminology.

Test Questions

1. The culture of Excalibur Semiconductors IT group emphasizes rapid response and resolution to crises and problems. This has led to a more reactive style of management rather than proactive, and is also known as the_______________.

   A. Action- Reaction Dynamic

   B. Strategic Response

   C. Capability Trap

   D. Fire-fighter Trap

2. The liaison between the finance team requesting additional application functionality and the application's end users is the _______________.

   A. Supervisor

   B. Power user

   C. Advisor

   D. Super user

3. Worldwide Aircraft implemented a new ERP application for one division to streamline its procurement efforts. The application was so successful the division wanted to add finance and human resource capabilities to the platform. The IT group responded by deploying those additional capabilities. This is an example of ____________________.

   A. A strategic asset

   B. A service asset.

   C. Service management as a closed-loop control system

   D. Attaining a higher service level

A. Reactive

B. Decisive

C. Proactive

D. Strategic

4. __________________ is an event that interrupts or has the potential to interrupt service.

A. Catastrophe

B. Outage

C. Failure

D. Incident

5. Service Transition provides value for business by decreasing the difference between __________ and __________.

A. Contracts, Re-negotiations

B. Budgeted costs, Actual costs

C. Systems, Configurations

D. Changes, Updates

6. Setting an annual budget for the cost of improving existing services is an often overlooked planning process. By having funding in place, the Continual Service Improvement team can take ___________ action to trends discovered through analysis of performance metrics.

7. Fill in the blank: ________ and ________ are used to determine priority.

A. Urgency and cost

B. Impact and cost

C. Urgency and impact

D. Capacity and demand

8. Utility is to __________ as warranty is to _______________.

A. Purpose, Insurance

B. Function, Attributes

C. Profit, Loss

D. Value, Repair

9. Fill in the blank: Supplier contract details are stored in the ________________

A. Configuration management system

B. ERP management system

C. CI registry

D. Service knowledge management system

10. The output of the CSI planning phase is a ________________.

A. Critical Success Plan

B. Improvement Process

C. Metrics Database

D. Service Improvement Plan

11. Testing of a new service takes place during the ____________ Lifecycle phase?

A. Service Operation

B. Service Design

C. Service Transition

D. Service Availability

12. Goals and requirements should always be ______________.

A. RACI

B. PCDA

C. SMART

D. RTO

13. Deployment of a service where none existed before is called ______________?

A. A brownfield deployment

B. A greenfield deployment

C. Risk-avoidant

D. Non-impacting

14. The release of a new or changed service into production is governed by ______________.

A. Service Transition

B. Service Management

C. Configuration Management

D. Service Delivery

15. __________ is the core of the ITIL Service Lifecycle.

A. Service Strategy

B. Service Operation

C. Service Design

D. Continual Service Improvement

16. _______________ is the first point of contact for IT service users, and is responsible for processing incidents and requests.

A. Service Staff

B. Service Desk

C. Help Desk

D. Support Staff

17. Fill in the blank: The Service Catalog is a subset of the

__________________.

A. Service Portfolio

B. Service Offering

C. Service Pipeline

D. Service Strategy

18. __________________ are a necessary condition for a good result in a service or process, and are typically measured by KPIs.

A. Uptime Factors

B. Critical Success Factors

C. System Metrics

D. Management Input Opportunities

19. Capital, infrastructure, applications, and information are examples of ___________ .

A. Supplies

B. Resources

C. Capacity

D. Capabilities

20. A roll back plan and remediation criteria are decision support tools provided during the ____________ phase.

A. Service Transition

B. Service Delivery

C. Early Life Support

D. Change Control

ITIL Practice Test: Fill in the Blank Answer Key and Explanations

1. D - The Fire-fighter Trap results when an organization too frequently rewards managers for putting out fires quickly, which can cause performance to suffer over the long run. ITIL - Key Principles and Models [ITIL - Key Principles and Models]

2. D - Super users are business users who act as liaisons between business and IT. ITIL - Selected Roles [ITIL - Selected Roles]

3. C - Services stem from service assets. Service potential is converted into performance potential for customer assets. Enhancing the performance potential often stimulates the request for additional services in terms of scale or scope. This request translates into more use of the service assets and justifies the retention of maintenance activities and upgrades. From this perspective, service management is a closed-loop control system. ITIL - Selected Processes [ITIL - Selected Processes]

4. D - An incident is any event that interrupts or can interrupt a service. Events may be reported by customers, the service desk, or a tool. ITIL - Generic Concepts and Definitions [ITIL - Generic Concepts and Definitions]

5. B - An effective Service Transition ensures that new or changed services are better aligned with the customer's business operation, including less deviation between planned budgets and actual costs. ITIL - Service Lifecycle [ITIL - Service Lifecycle]

6. C - It is recommended that an annual budget is set for the Service Improvement Plan, to permit quick actions to mitigate potential service incidents, rather than reacting to service outages that have already occurred. ITIL - Selected Processes [ITIL - Selected Processes]

7. C - Urgency and impact are used to determine priority. [The Service Lifecycle]

8. C - Utility represents the increase of a possible profit. Warranty represents the decline in possible losses. ITIL - Key Principles and Models [ITIL - Key Principles and Models]

9. D - The service knowledge management system (SKMS) contains all supplier and contract details, including the types of services or

commodities provided by each supplier. [ITIL Selected Processes]

10. D - The 7-step CSO improvement process describes how a Continual Service Improvement team should measure and report on the services in the Service Catalog. The planning phase for CSI yields a Service Improvement Plan (SIP). ITIL - Selected Processes [ITIL - Selected Processes]

11. C - New or enhanced services are tested to ensure proper functionality prior to release during the Service Transition phase. [The Service Lifecycle]

12. C - SMART is an acronym for Specific, Measurable, Achievable/ Appropriate, Realistic/ Relevant, and Timely/ Timebound. ITIL - Generic Concepts and Definitions [ITIL - Generic Concepts and Definitions]

13. 13.) B - Implementation of a Service where none existed before is called a greenfield deployment. A brownfield deployment is an upgrade or change to an existing deployment. ITIL - Generic Concepts and Definitions [ITIL - Generic Concepts and Definitions]

14. A - Service Transition includes the management and coordination of the processes, systems, and functions required for the building, testing, and deployment of a release into production. ITIL - Service Lifecycle [ITIL - Service Lifecycle]

15. A - Service Strategy is the axis of the Service Lifecycle that "runs" all other phases. ITIL - Service Lifecycle [ITIL - Service Lifecycle]

16. B - A service desk is a functional unit with a number of staff members who deal with a variety of service events. It must be the prime contact point for IT users. ITIL - Selected Roles [ITIL - Selected Roles]

17. A - The Service Portfolio represents the opportunities and readiness of a service provider to serve their customers and the market. The Service Portfolio can be divided into three subsets of services: the Service Catalog, the Service Pipeline, and Retired Services. ITIL - Service Lifecycle [ITIL - Service Lifecycle]

18. B - A critical success factor is an element that is crucial to an organization's ability to achieve a specific goal. ITIL - Selected Processes [ITIL - Selected Processes]

19. B - Resources are the direct input for production, and together with capabilities form the basis for the value of a service. ITIL - Generic Concepts and Definitions [ITIL - Generic Concepts and Definitions]

20. A - Service Transition develops systems and processes for knowledge transfer as necessary for an effective delivery of the service, and to make organization and support decision-making possible. ITIL - Selected Functions [ITIL - Selected Functions]

Knowledge Area Quiz: Generic Concepts and Definitions Practice Questions

Test Name: Knowledge Area Quiz: Generic Concepts and Definitions
Total Questions: 10
Correct Answers Needed to Pass: 7 (70.00%)
Time Allowed: 10 Minutes

Test Description

This practice test focuses on ITIL Generic Concepts and Definitions.

Test Questions

1. The management team at a large data center has just been informed by their telecommunications provider that a fiber cable cut has caused a regional outage that will not be repaired for at least 12 hours. This duration exceeds the SLAs for the data center, and the management team notifies the executive management team of this situation. The executive management team issues communications to affected business units regarding the situation and the potential duration of the outage. What is this type of escalation called?

   A. Hierarchical Escalation

   B. Priority Escalation

   C. Functional Escalation

   D. 4th Tier

2. What is a collection of authorized changes to a service called?

   A. Change Control

   B. Configuration Items

   C. Release

   D. Deployment packages

3. Configuration Items and the relationships between them are captured in which repository?

   A. Service Asset Database

   B. Change Management Database

   C. Configuration Management Database

   D. Relational Database

4. The network engineering team at Atlantis Communications is comparing results of tests conducted on a network that has been recently upgraded with what are considered to be the standard performance levels. What is the standard called?

   A. Baseline

   B. Change Point

   C. Performance Report

   D. Service Level

5. What is the primary source of demand for services?

   A. Incentives

   B. Customers

   C. Business Processes

   D. Activities

6. After conducting a Business Impact Analysis on the Balfour Glassware Corporation's IT infrastructure, it is determined that in the event of a disaster, certain file servers can be brought back up on line several days after the rest of the organizations' more critical assets. What is this type of recovery called?

   A. Gradual Recovery

   B. Intermediate Recovery

   C. Interim Recovery

   D. Manual Recovery

7. What type of change follows a pre-existing procedure or work instruction, and does not require an RFC?

   A. Normal

   B. Minor

   C. Emergency

   D. Standard

8. Which of the following is a downtime measurement of a system or a service?

   A. MTRS

   B. MTBSI

   C. MTSB

   D. MTBF

9. A customer information database is guaranteed to be available Monday-Friday, 8:00 AM - 5:00 PM. What is this guarantee called?

A. Warranty

B. Value

C. Utility

D. Underpinning Contract

10. Incidents are categorized by what two characteristics?

A. Severity and criticality

B. Resolution and recovery

C. Urgency and impact

D. Investigation and Diagnosis

Knowledge Area Quiz: Generic Concepts and Definitions Answer Key and Explanations

1. A - Hierarchical escalations, also known as vertical escalations are used when resolution of an incident will not be within set timeframes or be a satisfactory resolution. [Service Operation] [ITIL - Generic Concepts and Definitions]

2. C - A group of authorized changes to a service is called a release. [Service Transition] [ITIL - Generic Concepts and Definitions]

3. C - The CMDB is a set of one or more connected databases and information sources that provide a logical model of the IT infrastructure, including CIs and their relationships to each other. [Service Transition] [ITIL - Generic Concepts and Definitions]

4. A - A baseline is a benchmark used as a reference for later comparison. [Continual Service Improvement] [ITIL - Generic Concepts and Definitions]

5. C - Business processes are the primary source of demand for services. Patterns of business activities (BPA) influence the demand patterns seen by the service providers. [Service Strategy] [ITIL - Generic Concepts and Definitions]

6. A - A gradual recovery from a disaster offers recovery times in terms of days, rather than hours. Systems and services are returned to service in >72 hours. [Service Design] [ITIL - Generic Concepts and Definitions]

7. D - A standard change is a pre-approved change that is low risk, relatively common, and follows a procedure or work instruction These types of changes include password resets or providing a replacement keyboard. [Service Transition] [ITIL - Generic Concepts and Definitions]

8. A - MTRS, or Mean Time to Restore Service, is a measure of the average time a service or system is down after a failure. [Service Design] [ITIL - Generic Concepts and Definitions]

9. A - Warranty is the assurance that a service will meet specific requirements for availability, capacity, and reliability. A warranty is sometimes anchored with a formal agreement such as a Service Level Agreement. [ITIL - Generic Concepts and Definitions]

10. C - Incidents are prioritized by impact and urgency. [Service Operation]

[ITIL - Generic Concepts and Definitions]

Knowledge Area Quiz: Generic Concepts and Definitions - Answer Key and Explanations

Knowledge Area Quiz: Selected Processes Practice Questions

Test Name: Knowledge Area Quiz: Selected Processes
Total Questions: 10
Correct Answers Needed to Pass: 7 (70.00%)
Time Allowed: 10 Minutes

Test Description

This practice test focuses on additional processes of ITIL.

Test Questions

1. John has been named the Configuration Manager for InfoTech Systems. Part of his job is to define the naming convention for Configuration Items. What types of items should he consider as he develops this naming convention?

   A. System config files

   B. Documentation

   C. Batch jobs

   D. All of the above

2. Monitoring utilization of CPU, disk space and network with the intent to proactively increase the amount of these resources available for use is called:

   A. Component Capacity Management

   B. Bottlenecking

   C. New Product Introduction

   D. Service Capacity Management

3. High Seas Semiconductor has a very mature shared services IT group that services the IT needs of all the organizations' locations and subsidiaries around the world. What is the term for the aggregation of data and tools that High Seas uses to manage the full lifecycle of its services?

   A. Configuration Management Database

   B. Configuration Management System

   C. Service Knowledge Management System

   D. Service Catalog

4. Which of the following is an example of an event?

   A. Firewall failure

   B. Reduction in application response time

   C. A hard drive reaches its maximum capacity

   D. All of the above

5. A problem with an accounting application has been discovered by the finance team. The application support group has not identified a cause or resolution for this problem, but has devised a workaround for the finance team to continue working until a permanent resolution is found. This is an example of what type of service response?

   A. Incident Management

   B. Problem Management

   C. Change Management

   D. Event Management

6. Which Service Design process attempts to find the balance between resources, capabilities, and demand?

   A. Service Level Management

   B. Supplier Management

   C. Capacity Management

   D. Financial Management for IT Services

7. When should a Configuration Audit take place?

   A. After the detection of an unauthorized change.

   B. Following the recovery from a disaster.

   C. Before and after major changes to the IT Infrastructure.

   D. All of the above

8. The IT team at Super Shredder is having difficulty getting value out of the Knowledge Management System they have implemented. Which of the following is NOT a challenge to successfully implementing and using a Knowledge Management System?

A. The database is created by the Problem Management team but is used by both the Incident and Problem Management teams

B. Getting staff to use the system

C. Having time to record relevant information and knowledge after the actions are made.

D. Managing information that is no longer correct or relevant for the organization.

9. What type of study should be done to determine the ramifications of a specific event at an organization?

A. Return on Investment

B. Business Impact Analysis

C. Gap Analysis

D. PDCA

10. Florida Freight and Trucking is conducting a review of its operations to identify vital business functions, underlying dependencies, and recovery objectives. What is this activity called?

A. Risk analysis

B. Restore point objectives

C. Business impact analysis

D. Risk management

Knowledge Area Quiz: Selected Processes Answer Key and Explanations

1. D - The CMDB contains all types of data and information, including operating system files, hardware information, batch jobs, and operating instructions. The CMDB naming convention must be flexible enough to handle a variety of data and information types as a result. [Service Transition] [ITIL - Selected Processes]

2. A - Component Capacity Management identifies and manages each component in the IT Infrastructure, monitoring for bottlenecks, under or over utilization of resources, and repositioning or adding additional resources as needed Examples include upgrading a saturated WAN circuit from DS3 to 10 gigabit optical Ethernet, or routing specific traffic to a relatively underutilized Internet connection. [Service Design] [ITIL - Selected Processes]

3. C - The Service Knowledge Management System (SKMS) is the complete set of tools and databases that are used to manage knowledge and information necessary to make decisions and execute IT Service Management processes. The SKMS includes the Configuration Management System, the Configuration Management Database, the Known Error Database, and other relevant service data repositories and tools. [Service Transition] [ITIL - Selected Processes]

4. D - All of the above. An event is any change to the state of a configuration item that has some significance for the management of that CI. Events include, but are not limited to, total systems failure, a component exceeding its maximum performance threshold, or an item that has changed in a way that could impact service delivery. It is important to note that an event does not always result in a service interruption. [ITIL - Selected Processes]

5. A - The goal of incident management is to restore normal service operation as quickly as possible to minimize impact on the customer. As such, it addresses the symptoms but does not deal with the root cause. [Service Operation] [ITIL - Selected Processes]

6. C - The goal of Capacity Management is to ensure the current and future capacity and performance demands of the customer regarding IT service provision are delivered against justifiable costs. [Service Design] [ITIL - Selected Processes]

7. D - Reviews and audits verify the existence of configuration items, checking that they are correctly recorded in the Configuration Management Database, and that there is conformity between the documented baselines and the actual environment to which they refer. [Service Transition] [ITIL - Selected Processes]

8. A - The design and use of a Knowledge Management System is not restricted to any specific group of users. It may be used by all staff that require access to organizational knowledge in order to make service related decisions. [Service Transition] [ITIL - Selected Processes]

9. B - A Business Impact Analysis (BIA) studies a specific event, such as a power outage or flood, and the possible results the event could have on a company. [Service Design] [ITIL - Selected Processes]

10. C - During business impact analysis an organization identifies the functions that are absolutely critical to the success of the organization, the underlying services, infrastructure, and other dependencies required to support them, and how quickly and to what extent these functions must be brought back online in the event of an outage. [ITIL - Selected Processes]

ITIL Foundation
Practice Exam 9
Practice Questions

Test Name: ITIL Foundation Practice Exam 9
Total Questions: 40
Correct Answers Needed to Pass: 30 (75.00%)
Time Allowed: 60 Minutes

Test Description

This is the ninth cumulative ITIL Foundation test which can be used as an indicator for overall performance. This practice test includes questions from key ITIL areas.

Test Questions

1. The Availability Manager is tasked with ensuring adequate availability of all IT Services. How is adequate availability defined?

   A. No outages during core business hours

   B. Availability matches or exceeds the business requirements

   C. 100% availability

   D. 99.9% availability during core business hours

2. Which of the following is a model for defining the roles of stakeholders in a process or an activity?

   A. RACI

   B. Balanced Score Card

   C. SMART

   D. PDCA

3. Detection and repression are activities that occur at what point in the security enforcement model?

   A. Between control and evaluation

   B. Between threat and incident.

   C. Between damage and control

   D. Between incident and damage.

4. A service outage has occurred at Red River Systems. The outage had minimal impact on mission critical services, but it did hinder the ability of many users to access files needed for week-ending processes. The root cause was discovered and a plan was

put in place to correct the underlying problem and ensure that it does not happen again. What is this plan called?

A. Service Improvement Plan

B. Configuration Management Plan

C. Business Service Continuity Plan

D. Service Level Plan

5. The document that describes how a specific release will be handled is called what?

A. CMDB

B. Service Catalog

C. Rollout Plan

D. Release Management

6. Which of the following contain the customer view of IT services?

A. Business Service Catalog

B. System Catalog

C. Configuration Catalog

D. Technical Service Catalog

7. Information Security Management maintains which of the following key information properties?

A. Confidentiality, integrity, uptime

B. Confidentiality, integrity, availability

C. Configuration, integrity, accuracy

D. Confidentiality, intelligence, availability

8. In which lifecycle phase is the request fulfillment process executed?

A. Continual Service Improvement

B. Service Desk

C. Service Operation

D. Service Transition

9. IT Operations Management is usually divided into which two groups?

A. Application Management and Network Management

B. Desktop Management and Infrastructure Management

C. IT Operations Control and Facilities Management

D. Service Level Management and Technical Management

10. The ITIL Service Lifecycle has five phases. Which of the following best describes the relationship between each of the phases?

A. The phases are executed in a specific order.

B. Each phase is executed only once.

C. Each phase can affect any other phase.

D. Each phase affects only the following phase.

11. Charging customers a specific dollar amount for services received is standard operating procedure with commercial service providers. In shared service arrangements, where the provider and the customer may be part of the same company, organization, or division, direct payment of costs by customer to service provider is not as common. What is one way of handling this type of arrangement?

A. Indirect costs

B. RFC management

C. Accrual accounting

D. Notional charging

12. A minor normal change would be authorized by which of the following roles?

A. Change Advisory Board

B. Shift Manager

C. IT Management Board

D. Change Manager

13. Seven Seas Semiconductor uses Service Desks in different time zones to support the round-the-clock needs of their global organization. This is an example of what type of Service Desk?

A. Follow-the-Sun

B. On Demand

C. Local

D. Virtual

14. Which of the following is not a reactive activity?

A. Incident Handling

B. Problem Resolution

C. Monitoring

D. Capacity Planning

15. Posting a patch or updated driver on a website where users can download and install the software at their discretion is known as what type of release?

A. Phased Approach

B. Pull Approach

C. Manual Approach

D. Push Approach

16. What type of analysis must be conducted before implementing a security measure?

A. ROI

B. SWOT

C. RACI

D. Cost- Benefit Analysis

17. An online delivery tracking tool is an example of what type of capability?

A. Automated process

B. Status reporting

C. Self-help

D. Request fulfillment

18. All but one of the following are Service Operation Processes. Which is the incorrect process?

A. Access Management

B. Event Management

C. Request Management

D. Configuration Management

19. Which of the following terms refers to the ability of a service or component to perform as intended over a specific duration of time?

A. Resilience

B. Reliability

C. Availability

D. Serviceability

D. Change Advisory Manager and Service Operation Manager

20. The Business Continuity team at Prendergast Manufacturing has begun evaluating its assets, and the threats to and vulnerabilities of these assets. What is this process called?

A. Risk Assessment

B. Daily Operational Activity

C. Business Impact Analysis

D. Risk-avoidant behavior

21. MaxStorage has recently reorganized its IT structure and implemented a Technical Management team. This team helps plan, implement, and maintain a stable technical infrastructure to support the organization's business processes. What roles will be included in this team?

A. Senior and Line Managers

B. CIO and CTO

C. Specialist Technical Architects and Designers and Specialist Maintenance and Support Staff

22. A large aircraft engineering corporation has created a repository containing information about its shared data integration middleware service. It includes information about the entire lifecycle of the service, from business and service level requirements to operations plans and acceptance criteria. What is this collective body of information called?

A. Service Design Package

B. Service Catalog

C. Business Impact Analysis

D. Service Portfolio

23. Service Demand can be controlled by which of the following mechanisms?

A. Service Catalog improvements and Business Impact Analysis

B. Effective IT governance and auditing processes.

C. Management policies, corporate directives, and dollar cost averaging

D. Physical/ technical constraints and financial constraints

24. Which role is responsible for managing Underpinning Contracts?

A. Service Design Manager

B. Supplier Manager

C. Service Operation Manager

D. Service Strategy Manager

25. In which lifecycle phases is Supplier Management carried out?

A. Service Design, Service Delivery, and Continual Service Improvement

B. Service Design, Service Operation, and Continual Service Improvement

C. Service Strategy, Service Transition, Service Operation

D. Service Design, Service Operation, and Contract Management

26. Jonathan is the manager of a desktop support group that deals with a number of hardware vendors which provide new equipment and replacement components. Where would he record agreements between the suppliers and his company regarding service improvements?

A. Service Level Agreement

B. Service Catalog

C. Supplier Contract Performance Report

D. Supplier Service Improvement Plans

27. Hill and Meyer, LLC have purchased a 3rd party maintenance contract for 8x5x4 onsite support for specific network components to limit potential downtime in the event of an outage. What is this type of contract called?

A. Service Level Agreement

B. Service Level Agreement

C. Operational Level Agreement

D. Underpinning Contract

28. The finance division at a large manufacturing company runs a large reporting batch job starting at 6pm on the last business day of the month. This job must be completed by 8am the next business day. Staggering or delaying the start of backups on the servers and storage devices used during these jobs to ensure there is no slowdown in the month-end reporting process is what type of activity?

A. Demand Management

B. Service Level Management

C. Business Process Alignment

D. Operational Level Agreement

29. Service Pipeline, Service Catalog, and Retired Services are defined in which of the following?

A. Service Offering

B. Service Portfolio

C. CMDB

D. Service Library

30. Demand Management is found in which phase?

A. Service Operation

B. Service Strategy

C. Continual Service Improvement

D. Service Design

31. During which phase of the Service Lifecycle are IT budgets set?

A. Service Operation

B. Service Transition

C. Service Design

D. Service Strategy

32. Which role is responsible for ensuring SLA targets for incident resolution are met?

A. Service Desk

B. Event Manager

C. Incident Manager

D. Service Operation Manager

33. Which of the following is NOT a core ITIL volume?

A. Service Delivery

B. Service Transition

C. Service Design

D. Service Operation

34. Service Level Management processes are found within which two lifecycle phases?

A. Service Design and Service Operation

B. Service Transition and Service Operation

C. Service Operation and Continual Service Improvement

D. Service Design and Continual Service Improvement

35. Two companies have decided it would be in their best interests to jointly create disaster recovery (DR) plans that will allow one company to co-locate their mission critical gear in the data center of the other in the event one of their data centers suffers a disaster-level outage. What is this DR approach called?

A. Gradual Recovery

B. Counter Measures

C. Reciprocal Arrangement

D. Shared Services

36. Sonata Hosting offers its customers a basic hosting package, along with several optional supporting services such as additional disk space and static IPs, as well as enhanced service level add-ons offering faster download speeds. What is the comprehensive offering called?

A. Excitement Factors

B. Service Provider

C. Service Package

D. Shared Services

37. Which of the following is not a Service Operation Function?

A. Service Desk

B. Technical Management

C. Application Management

D. Event Management

38. The Information Security Manager at Higgins and Clark has just

implemented a formal control process. Which of the following security models shows the correct location of the control process?

A. Threat, Control, Incident, Damage

B. Threat, Incident, Damage, Control

C. Control, Threat, Incident, Damage

D. Threat, Incident, Control, Damage

39. What perspectives must be considered when developing an Information Security Management approach?

A. Organizational, procedural, physical, and technical

B. Disk, network, buffer overflows, spoofing

C. Threats, vulnerabilities, hacks, and intrusions

D. Physical, logical, data, and infrastructure

40. Which of the following types of Service Desks has a low first-contact resolution rate by design?

A. Central Service Desk

B. Call Center

C. Help Desk

D. Virtual Service Desk

ITIL Foundation Practice Exam 9 Answer Key and Explanations

1. B - The Availability Manager seeks to deliver availability that matches or exceeds business requirements. [Service Design] [ITIL - Selected Roles]

2. A - The RACI model helps teams identify who is Responsible, Accountable, Consulted, and Informed. [Service Design] [ITIL - Key Principles and Models]

3. D - The security enforcement lifecycle model is Threat, Incident, Damage, Control. In the event a security incident occurs, it must first be detected; subsequent measures to repress or minimize the damage are then taken. [Service Design] [ITIL - Key Principles and Models]

4. A - Service Improvement Plans are formal plans to implement improvements to a process or service. They are used to ensure that improvement actions are identified and carried out on a regular basis. [Service Operation] [ITIL - Key Principles and Models]

5. C - The Rollout Plan is the documented approach for distributing a single release. [Service Transition] [ITIL - Service Lifecycle]

6. A - The Business Service Catalog contains details of all the IT services delivered to the customer, along with the relationships to the business units and the business processes that rely on the services. [Service Design] [ITIL - Selected Functions]

7. B - Information Security Management ensures that the confidentiality, integrity, and availability of an organization's assets are maintained. [Service Design] [ITIL - Selected Functions]

8. C - The Service Operation Phase is concerned with the execution of strategies, designs, and plans from all Service Lifecycle phases, with primary focus on these processes: Incident Management, Problem Management, Event Management, Request Fulfillment, and Access Management. [Service Operation] [ITIL - Service Lifecycle]

9. C - IT Operations Management is usually divided into IT Operations control, which ensures that routine operational tasks are carried out, and Facilities Management which is charged with the management of the

physical IT environment such as data centers and computer rooms. [Service Operation] [ITIL - Service Lifecycle]

10. C - Each of the ITIL Service Lifecycle phases can affect any other phase. Further, there is no specific order in which the phases are executed, and each phase may be executed continuously or repeatedly. [The Service Lifecycle]

11. D - In notional charging, the costs incurred in the provision of a specific service are communicated to the recipient, but no payment is required. This method is useful for encouraging more efficient use of IT resources. [Service Strategy] [ITIL - Service Lifecycle]

12. D - Normal changes are non-emergency changes. Minor normal changes are handled by the Change Manager. Significant normal changes are approved by the Change Advisory Board. Major normal changes are approved by the IT Management Board. [Service Transition] [ITIL - Selected Roles]

13. A - The Follow-the-Sun Service Desk approach establishes multiple Service Desks in different time zones to provide 24x7 support. [Service Operation] [ITIL - Key Principles and Models]

14. D - Reactive activities are those that revolve around detecting and handling events that have already occurred, and include activities such as monitoring, problem and incident management, and outages. [Service Design] [ITIL - Selected Processes]

15. B - The Pull Approach is characterized by the posting of a software release in a central location for users to download and install on their own timetable.[Service Transition] [ITIL - Selected Processes]

16. D - A cost-benefit analysis must be conducted before implementing a security measure, to ensure the measure is appropriate for both the threat risk and potential loss.[Service Design] [ITIL - Selected Processes]

17. C - Self-help capabilities allow users to perform frequently requested tasks themselves. These tasks can include online delivery tracking systems, self service password changes, and IVR systems that provide account information. These systems reduce the amount of calls received by a Service Desk and increase the efficiency of the service delivery. [Service Operation] [ITIL - Selected Processes]

18. D - The processes that support Service Operation are Event Management, Incident Management, Problem Management, Request Management, and Access Management. [Service Operation] [ITIL - Selected Processes]

19. C - Availability is the ability of an IT service or component to perform its required function at a stated instant or over a stated period of time. [Service Design] [ITIL - Selected Processes]

20. A - Risk assessment is the evaluation of an organization's assets, threats, and vulnerabilities. [Service Design] [ITIL - Selected Processes]

21. C - It is important that the Technical Management function be made up of both support and design staff to ensure that a quality, supportable design is implemented. [Service Operation] [ITIL - Selected Roles]

22. A - A Service Design Package includes all aspects of a service and its requirements, and is used to provide guidance and structure through the entire lifecycle of the service. [Service Design] [ITIL - Generic Concepts and Definitions]

23. D - Physical/ Technical Constraints such as bandwidth throttling and session timeouts, and Financial Constraints such as penalties for usage in excess of base levels are methods for controlling demand and usage of services. [Service Strategy] [ITIL - Selected Processes]

24. B - The Supplier Manager is responsible for managing Underpinning Contracts, as it deals with third-party service providers on which service deliverables for the SLA have been built. [Continual Service Improvement] [ITIL - Selected Roles]

25. B - Supplier Management is carried out in the Service Design, Service Operation, and Continual Service Improvement phases. [Service Design] [ITIL - Service Lifecycle]

26. D - Supplier Service Improvement Plans are used to record all improvement actions and plans agreed upon between suppliers and service providers. [Service Design] [ITIL - Selected Processes]

27. D - An underpinning contract is between an organization and an external supplier that supports the IT organization in the delivery of services. [Service Design] [ITIL - Selected Processes]

28. A - The goal of Demand Management is to assist the IT service provider in understanding and influencing customer demand for services and the provision of capacity to meet these demands. Demand Management techniques include prioritizing critical reports and batch jobs and staggering the start of less critical activities that could utilize resources needed for the higher priority activities. [Service Strategy] [ITIL - Selected Processes]

29. B - A Service Portfolio includes the complete set of services managed by a Service Provider, and includes the Service Pipeline, the Service Catalog, and Retired Services. [Service Strategy] [ITIL - Service Lifecycle]

30. B - Demand Management has been made a separate process within Service Strategy in ITIL. [Service Strategy] [ITIL - Selected Processes]

31. D - IT budgets are set as part of the Service Strategy phase. [Service Strategy] [ITIL - Service Lifecycle]

32. C - The Incident Manager is responsible for the management of the Incident Management Team, the effectiveness and efficiency of that team, and meeting or exceeding SLAs for incident resolution. [Service Operation] [ITIL - Selected Roles]

33. A - The ITIL Library has 3 core volumes: Service Strategy, Service Design, Service Transition, Service Operation, and Continual Service Improvement [ITIL Overview] [ITIL - Service Management as a Practice]

34. D - Service Level Management is found in two Service Lifecycle phases: Service Design and Continual Service Improvement. [Service Design] [ITIL - Selected Processes]

35. C - Two organizations that plan to utilize the resources of the other in the event of a disaster for one of the organizations have entered into what is called a reciprocal arrangement. [Service Design] [ITIL - Key Principles and Models]

36. C - A Service Package provides a detailed description of a service to be delivered to customers. The contents include the core service, supporting services, and service level packages. [Service Strategy] [ITIL - Selected Functions]

37. D - Service Desk, Technical Management, Application Management, and IT Operations Management are Service Operation functions. Event Management is a Service Operation process. [Service Operation] [ITIL - Selected Functions]

38. B - The Security Process Model is Threat, Incident, Damage, Control.

[Service Design] [ITIL - Selected Processes]

39. A - Information Security Management ensures that the confidentiality, integrity, and availability of an organization's assets are maintained. Information Security Management must consider the following four perspectives: Organizational, Procedural, Physical, and Technical. [Service Design] [ITIL - Selected Processes]

40. B - Call Centers are intended to handle and log large volumes of calls, and typically have low resolution rates on the first customer contact. Calls are usually routed to more specialized staff for resolution. [Service Operation] [ITIL - Selected Roles]

Knowledge Area Quiz: Service Lifecycle Practice Questions

Test Name: Knowledge Area Quiz: Service Lifecycle
Total Questions: 10
Correct Answers Needed to Pass: 7 (70.00%)
Time Allowed: 10 Minutes

Test Description

This practice test specifically targets the ITIL concepts related to the Service Lifecycle.

Test Questions

1. Continual Service Improvement metrics fall into which three categories?

   A. KPIs, Activity Metrics, Component Metrics

   B. Performance Metrics, Process Metrics, Service Metrics

   C. Infrastructure Metrics, Process Metrics, Resource Metrics

   D. Technology Metrics, Process Metrics, Service Metrics

2. Knowledge Management is often visualized through what structure?

   A. Configuration Management Database

   B. RACI-Responsible, Accountable, Consulted, Informed

   C. DIKW- Data, Information, Knowledge, Wisdom

   D. Knowledge Management Database

3. Aristotle Publishing has decided to implement a shared services model for their IT infrastructure. In which Lifecycle phase will the shared services team design ask the question "Why do we need this Service?"

   A. Service Design

   B. Service Transition

   C. Service Strategy

   D. Continual Service Improvement

4. A fire in the datacenter that takes out several mission critical systems and leaves the company unable to conduct

key business work is an example of what type of Incident?

A. Catastrophic Incident

B. Managed Incident

C. Major Incident

D. Emergency Incident

5. South Seas Shipping IT division supports 17 business offices around the globe. The services offered and managed by the IT division, as well as the value proposition and business cases for the services are captured in what repository?

A. Service Portfolio

B. Service Description

C. Service Catalog

D. Service Package

6. Service Asset and Configuration Management is a key process in which Lifecycle Phase?

A. Service Strategy

B. Service Transition

C. Service Operation

D. Service Design

7. John is tracking the changing usage of a critical service and has discovered a significant increase in use over the last 2 weeks. He has created a report on these findings for the IT executives, who will evaluate the situation and determine if the issue is a matter of changing business requirements or a capacity issue. In which Lifecycle Phase is this work conducted?

A. Service Design

B. Service Operation

C. Service Transition

D. Continual Service Improvement

8. Strategic objectives are converted into services during which Lifecycle Phase?

A. Service Transition

B. Service Strategy

C. Service Operation

D. Service Design

9. The IT management team has developed a Release Policy to govern how all changes to a specific service are to be implemented. In what Lifecycle Phase is this policy developed?

   A. Service Operation

   B. Service Strategy

   C. Service Design

   D. Service Transition

10. John is part of an engineering and architecture team designing the network infrastructure for a new service offering. In what Lifecycle Phase is this work being conducted?

   A. Service Operation

   B. Service Design

   C. Service Strategy

   D. Service Transition

Knowledge Area Quiz: Service Lifecycle Answer Key and Explanations

1. D - The goal of Service Measurement and Reporting is coordinating the design of metrics, data collection, and reporting activities from other processes and functions. The three types of metrics an organization will need to collect to support Continual Service Improvement processes as well as other activities are: technology metrics, process metrics, and service metrics. [Continual Service Improvement] [ITIL - Service Lifecycle]

2. C - The objectives of Knowledge Management include supporting the service provider in order to improve efficiency and quality of services and ensuring that the service provider's staff have adequate information available. Knowledge Management is often visualized as Data, Information, Knowledge, and Wisdom. [Service Transition] [ITIL - Service Lifecycle]

3. C - One of the major concepts developed within the Service Strategy phase is determining how to create service value. This ensures that, before rushing out to determine how to design a Service, the organization stops to ask why the service is needed. [Service Strategy] [ITIL - Service Lifecycle]

4. C - A Major Incident is the highest category of impact for an Incident. A Major Incident results in significant disruption to the business. [Service Operation] [ITIL - Service Lifecycle]

5. A - A Service Portfolio describes a provider's services in terms of business value. They include the complete set of services managed by a service provider, as well as supporting data such as requirements, value proposition, risks, and costs. [Service Strategy] [ITIL - Service Lifecycle]

6. B - The processes in the Service Transition phase are Knowledge Management, Service Asset and Configuration Management, Change Management, Release and Deployment Management, and Validation and Testing. [Service Transition] [ITIL - Service Lifecycle]

7. D - Continual Service Improvement is concerned with improving services and processes through constant monitoring, reporting, evaluation, and improvements. [Continual Service Improvement] [ITIL - Service Lifecycle]

8. D - One of the main objectives of the Service Design phase is to convert strategic objectives identified during the Service Strategy Phase into Services and Service Portfolios. [Service Design] [ITIL - Service Lifecycle]

9. D - The Release Policy sets the release guidelines, constraints, and limits for the organization, and is developed in the Service Transition Phase of the Lifecycle. [Service Transition] [ITIL - Service Lifecycle]

10. B - The infrastructure, processes, and support mechanisms needed to meet the availability requirements of the customer are developed during the Service Design Phase. [Service Design] [ITIL - Service Lifecycle]

ITIL Foundation
Practice Exam 10
Practice Questions

Test Name: ITIL Foundation Practice Exam 10
Total Questions: 40
Correct Answers Needed to Pass: 30 (75.00%)
Time Allowed: 60 Minutes

Test Description

This is the tenth cumulative ITIL Foundation test which can be used as an indicator for overall performance. This practice test includes questions from key ITIL areas.

Test Questions

1. Dan is the Availability Manager of Mid-West Telecom. What level of availability is he responsible for achieving?

A. 99.99%

B. 99.50%

C. 100%

D. Availability that meets or exceeds the business requirements.

2. Bayview Bank has received a number of customer complaints about wait times at each of its branches during between 11:00 AM and 1:00 PM. Following a detailed analysis of all teller transactions, the bank determined that 30% of customer interactions with tellers are to obtain copies of deposited items and other banking records. To reduce customer wait times to 3 minutes or less, the business team has decided to expand its automated services by providing customers with the ability to retrieve this information from the bank's website as well as from its ATMs. In which Lifecycle stage was this decision made?

A. Service Design

B. Service Value

C. Service Strategy

D. Business Strategy

3. John is presenting the final infrastructure design for a new sales database to the other members of the engineering team. During which Lifecycle stage are infrastructure designs developed?

A. Service Strategy

B. Service Design

C. Service Operation

D. Service Availability

4. The goal of IT Operations Management is to perform the daily activities needed to manage IT Infrastructure, according to the standards defined in the Service Design phase. What are the two distinct divisions in IT Operations Management functions?

   A. IT Operations Control, Facilities Management

   B. Maintenance, Prevention

   C. Hardware Management, Software Management

   D. Application Management, Infrastructure Management

5. Eastern Enterprises uses password-based access to servers containing employee records. This is an example of using a tool to protect which security tenet?

   A. Confidentiality

   B. Availability

   C. Secrecy

   D. Integrity

6. Which of the following Lifecycle phases are primarily proactive?

   A. Service Transition and Service Operation

   B. Service Design and Continual Service Improvement

   C. Service Transition and Service Operation

   D. Service Design and Service Transition

7. What are the fundamental activities for Financial Management for IT Services?

   A. Budgeting, Fiscal Responsibility, Funding

   B. Direct Costs, Cost Elements, and Cost Types

   C. Budgeting, IT Accounting, Charging

D. Charging, Budgeting, Direct Costs

8. Dan is researching and analyzing his company's assets and the threats and vulnerabilities those assets are facing. What is this type of analysis called?

A. Business Impact Analysis

B. Disaster Recovery Analysis

C. Business Continuity Analysis

D. Risk Assessment

9. ACME, Inc. is considering upgrading the servers that run its corporate email system and the human resources database. Due to budgetary constraints, however, only one system can be upgraded at this time. It is determined that the corporate email system will be upgraded now, and the human resources database will be upgraded at a later time. During which Lifecycle phase is this decision made?

A. Service Transition

B. Service Operation

C. Service Design

D. Continual Service Improvement

10. Acme Enterprises IT group has an SLA with all its internal customers guaranteeing that services will be available Monday-Friday, 8am-5pm. A team within the IT group constantly monitors all aspects of the service offering to be sure that the appropriate resources and infrastructure are in place to meet or exceed the terms of the SLA. What is the name of this process?

A. Availability Management

B. Service Analysis

C. Capacity Management

D. Service Reviews

11. Which process or function provides a way of comparing the actual performance of a service against its design and SLAs?

A. Financial Management

B. Demand Management

C. Event Management

D. Capacity Management

12. Which Service Transition process is responsible for managing information derived from a number of sources; such as user and support documentation?

A. Service Asset and Configuration Management

B. Knowledge Management

C. Release and Deployment Management

D. Change Management

13. Steve is the Service Level Manager of a managed hosting provider. How does he identify and manage improvements to services and processes as part of Continual Service Improvement?

A. Constant monitoring, reporting, evaluating, and improving.

B. Delegating authority to the Service Strategy Team

C. Conducting quarterly Service Capability Reviews

D. Updating the Service Catalog

14. What data point would be used as a starting point to measure the effect of a Service Improvement Plan?

A. Capacity Analysis

B. Configuration Database Entry

C. Baseline

D. Service Analysis

15. Spare hardware is stored, in a pre-configured or pre-assembled and ready to deploy condition, in which area?

A. Configuration Management Database

B. Definitive Media Library

C. Hot Spares

D. Definitive Spares

16. Informatic Systems has deployed a database that tracks all incidents and problems that have occurred, as well as information such as root cause and resolution. This system helps staff reach a faster diagnosis and resolution for future of incidents and problems. What is this database called?

A. Knowledgebase

B. Incident Log

C. Known Error Database

D. Configuration Management Database

17. Which components make up the Service Knowledge Management System (SKMS)?

A. Configuration Management System, Accounting System, Knowledgebase

B. Configuration Management System, Configuration Management Database, Knowledge Management Database

C. Configuration Control System, Configuration Management Database, Knowledge Management Database

D. Configuration Management System, Configuration Management Database, Known Error Database

18. The finance department at Faucets and Fixtures International wants a new accounting application with features that will meet new regulatory mandates for their business. They have asked the IT group to assist with this effort. The IT group analyzes the new application and conducts a study to determine what the resource requirements are for the application, and gathers the specific performance and usage requirements for the finance team. This information is used to determine how large the underlying IT infrastructure such as servers and network gear should be. What is this process called?

A. User Acceptance Testing

B. Application Sizing

C. Continual Service Improvement

D. Capacity Analysis

19. A service package consists of a Service Level Package and one or more:

A. SLAs and Warranties

B. Supporting Services and Enhanced Services

C. Customer Service Packages and Core Services

D. Core Services and Supporting Services

20. Which process balances capacity and performance demands with costs?

A. Service Management

B. Financial Management

C. Demand Management

D. Capacity Management

21. Bernice is the release and deployment manager Payment Services, Inc. She is responsible for the plan, build, test, and deployment of new services at her organization. Which of the following is not one of the phases of this process?

A. Release and deployment planning

B. Define functionality

C. Release build and test

D. Review and close

22. John and Sally are members of a team measuring service performance data related to the service offerings of Global Storage. What are four reasons why measurements would be taken?

A. Protect, Resolve, Insure, Support

B. Hardware, Software, Network, SAN

C. Finance, Legal, Technical, Functional

D. Validate, Direct, Justify, Intervene

23. Tonya is the service owner of a web-based application at a large company that provides income tax preparation services. During which Lifecycle phase will she focus most on staying within budget and on schedule for the release of this application?

A. Service Operation

B. Service Design

C. Service Strategy

D. Service Transition

24. Which if the following is an example of a reactive change?

A. Implementing additional bandwidth prior to a peak business cycle.

B. Replacing a failed switch.

C. Upgrading a server that is running at 80% capacity.

D. Consolidating servers to save money.

25. Resolving an outage as quickly and efficiently as possible falls within which Lifecycle phase?

A. Service Transition

B. Availability Management

C. Service Operation

D. Continual Service Improvement

26. MaxData has implemented a series of workarounds to keep critical month end batch jobs running as intended. These workarounds do not resolve the problem, rather they allow the business teams to return to work on their month-end reporting tasks as rapidly as possible. What is this approach called?

A. Incident Management

B. Problem Resolution

C. Incident Resolution

D. Problem Management

27. Sonic Industries has implemented a network storage service for the entire organization to use. The cost of this service to each department is determined on a per-megabyte-used basis. No actual funds change hands however; the costs are used simply for tracking and monitoring purposes, in an effort to encourage efficient use of resources. What is this financial management method called?

A. Indirect Costs

B. IT Accounting

C. Cost Units

D. Notional Charging

28. Janet manages a call center and is interviewing candidates for customer support representatives. Which of the following is the most important for this role?

A. Communication Skills

B. Call Center Application Knowledge

C. Technical Skills

D. Industry Knowledge

29. Which of the following items can not be captured in a database?

A. Information

B. Data

C. Knowledge

D. Wisdom

30. Information Security Management ensures that which elements of an organization's assets, information, data, and services are maintained?

A. Confidentiality, Integrity, Availability

B. Integrity, Confidentiality, and Access Control

C. Availability, Security, User Rights

D. Baselines, Infrastructure, Processes

31. Senior management has been analyzing the types of calls logged by the Help Desk at West Coast Manufacturing. A significant percentage of the calls were found to be related to forgotten passwords and requests for new software on users' workstations. It is determined that a webpage to let users reset their passwords and request new software will be a useful service to deploy company wide. What is this type of support called?

A. Self-Help

B. Intranet

C. Single Sign-on

D. Service Desk

32. Shelly is the Change Manager for DataStor. When reviewing requests for changes, she consults with specific departments or staff to ensure there is approval in which three areas?

A. Stakeholders, Service Owners, Deployment Managers

B. Legal, Accounting, Help Desk

C. Financial, Business, Technology

D. Functional, Technical, Support

33. A planned recovery from a disaster in >72 hours is called what?

A. Reciprocal

B. Manual

C. Intermediate

D. Gradual

34. Markpoint Systems is evaluating its server components in an effort to ensure its infrastructure is suitable for the availability levels it offers its clients. Which of the following uptime measurements would be applied to the hard drive of a server?

A. Mean Time Between Failures

B. Mean Time Between System Incidents

C. Mean Time To Failure

D. Mean Time To Restore Service

35. Great Outdoors Telecom has developed an SLA for a specific level of service for all their customers. What is this SLA called?

A. Multi-Level SLA

B. Hierarchical SLA

C. Service-Based SLA

D. Customer-Based SLA

36. Which of the following improvements offers a short-term benefit?

A. New tools

B. Training

C. Server upgrades

D. Network infrastructure changes

37. While upgrading their SAN, a large manufacturing conglomerate suffers a server failure. The deployment team immediately initiates the remediation plan which includes building new servers with a specific configuration and, later once service is restored, comparing them to the failed servers in order to learn more about the cause of the outage. What is the specific configuration called?

A. Portfolio Baseline

B. New Growth Baseline

C. Configuration Baseline

D. Backout Baseline

38. Which of the following is an example of a Change Management Key Performance Indicator?

A. Number of Emergency Changes

B. Number of RFCs Accepted/ Rejected

C. Number of Implemented Changes

D. All of the above

39. The finance department at ACME Widgets is asking the IT team for the business justification to purchase and implement a CMDB system. Currently they do not have a unified CMDB; data is stored on several different systems and maintained by different groups. Which of the following would be appropriate justifications?

A. Possible cost savings related to identifying duplicate services.

B. Data is available to all staff

C. One single system to support.

D. All of the above

40. Which of the following groups at Starlight, Inc. is responsible for negotiating SLAs and OLAs?

A. Operation Management

B. Service Level Management

C. Supplier Management

D. Contract Management

ITIL Foundation Practice Exam 10 Answer Key and Explanations

1. D - The Availability Manager does not seek to achieve 100% availability, but instead seeks to deliver availability that matches or exceeds the business requirements. [Service Design] [ITIL - Selected Roles]

2. C - During the Service Strategy stage, the need for services- as well as their relative importance, value, and costs- is determined. [The Service Lifecycle]

3. B - The infrastructure, processes, and resources required to support a new service are designed and developed during the Service Design stage. [The Service Lifecycle]

4. A - IT Operations has two unique functions, IT Operations Control, and Facilities Management. [Service Operation] [ITIL - Selected Functions]

5. A - Protecting information against unauthorized access, by means of tools such as access cards, firewalls, or passwords, ensures the confidentiality of that information. [Service Design] [ITIL - Service Lifecycle]

6. D - The Service Design and Service Transition stages are proactive in nature. Activities and processes during these stages involve the development of new or enhancement of existing services and their support mechanisms. Service Operation and Continual Service Improvement activities and processes are undertaken in response to events or data that indicate the availability or value of a service has been impacted. [ITIL - Service Lifecycle]

7. C - There are three fundamental activities for Financial Management for IT Services (FMIT): budgeting, IT accounting, and charging. [Service Strategy] [ITIL - Selected Processes]

8. D - Risk Assessment is the analysis of the value of assets, identification of threats to those assets, and evaluation of how vulnerable each asset is to the identified threats. [Service Design] [ITIL - Selected Processes]

9. C - Service Design uses a holistic approach to ensure IT services supporting business processes and services that have been identified as critical are prioritized over less important IT services. [The Service Lifecycle]

10. A - Availability Management is the process responsible for defining,

analyzing, planning, measuring, and improving all aspects of the Availability of IT services, ensuring that all infrastructure, processes, tools, and other resources are appropriate for the agreed upon SLA targets for availability. [Service Design] [ITIL - Service Lifecycle]

11. C - The goal of event management is to enable stability in IT Services Delivery and Support by monitoring all events that occur throughout the IT infrastructure to allow for normal service operation and to detect and escalate exceptions. It also provides a way to compare actual performance and behavior against design standards and SLAs. [Service Operation] [ITIL - Selected Processes]

12. B - The goal of the Knowledge Management process is to enable organizations to improve the quality of management decision making by ensuring that reliable information and data is available throughout the service lifecycle. [Service Transition Management] [ITIL - Selected Processes]

13. A - Service Level Management requires constant monitoring, reporting, evaluating, and improving of services. Through this ongoing effort, the Service Level Manager identifies and manages improvements to services and processes. [Continual Service Improvement] [ITIL - Selected Roles]

14. C - A baseline is a benchmark used as a reference point, such as the start of a cycle of change, used to measure the effect or performance of that change. [Continual Service Improvement] [ITIL - Service Lifecycle]

15. D - Definitive Spares is the physical storage of all spare IT components and assemblies maintained at the same configuration level as within the production environment. [Service Transition] [ITIL - Key Principles and Models]

16. C - The purpose of a Known Error Database is to store knowledge about incidents and problems and how they were remedied, so that a quicker diagnosis and solution can be found if further incidents and problems occur. [Service Operation] [ITIL - Key Principles and Models]

17. D - The Service Knowledge Management System (SKMS) is the complete set of tools and databases that are used to manage knowledge and information. The SKMS includes the Configuration Management System (CMS), the Configuration Management Database (CMDB), and the Known Error Database (KEDB);

as well as the Service Portfolio, AMIS, CMIS, SMIS, and the SCD.

18. B - Application Sizing determines the hardware or network capacity required to support new or modified applications and the predicted workload. [Service Design] [ITIL - Generic Concepts and Definitions]

19. D - A service package is a detailed description of an IT service that can be delivered to customers. A service package consists of a Service Level Package and one or more core services and supporting services. [Service Strategy] [ITIL - Service Lifecycle]

20. D - The goal of Capacity Management is to ensure that the current and future capacity and performance demands of the customer regarding IT service provision are delivered against justifiable costs. [Service Design] [ITIL - Selected Processes]

21. B - The four phases of the release and deployment process are: 1) release and deployment planning, 2) release build and test, 3) deployment, and 4) review and close. [ITIL Selected Processes]

22. D - The four most common reasons to measure are: validate prior decisions, direct future goals, justify continued support, and intervene in problem areas. [Continual Service Improvement] [ITIL - Selected Processes]

23. D - One of the objectives of the Service Transition phase is ensuring that services are delivered within the pre-established parameters for cost and timing, also known as budget and schedule. [The Service Lifecycle]

24. B - Reactive changes are implemented in response to an incident, problem, or event, while proactive changes are implemented in advance of a predicted need. [Service Transition] [ITIL - Technology and Architecture]

25. C - During the Service Operation phase an organization delivers and maintains a service to its customer. Minimizing the occurrence, impact, and duration of service outages is an important objective of this phase. [The Service Lifecycle]

26. A - The goal of incident management is the restoration of normal service operation as quickly as possible and minimizing the adverse impact on business operations, thus ensuring that the best possible levels of service quality are maintained. [Service Operation] [ITIL - Selected Processes]

27. D - Charging customers for their use of IT Services can be implemented in a number of ways in order to encourage more efficient use of IT resources. Notional charging is one option, in which the costs of providing Services to customers are communicated but no actual payment is required. [Service Strategy] [ITIL - Service Management as a Practice]

28. A - Communication is the most important skill for call center staff, as the primary role of the call center is to act as the single point of contact between the end users and the service provider. Staff must be able to deal with a wide range of people and situations. [Service Operation] [ITIL - Selected Roles]

29. D - Tools and databases can be used to capture Data, Information, and Knowledge, while Wisdom is a concept relating to abilities to use knowledge to make correct judgments and decisions. [Service Transition] [ITIL - Key Principles and Models]

30. A - Information Security Management ensures that the confidentiality, integrity, and availability of an organization's assets, information, data, and IT services are maintained. [Service Design] [ITIL - Selected Functions]

31. A - Many organizations find it beneficial to offer Self Help capabilities such as web pages to their users. This reduces the load on the Service Desk and can lead to greater efficiency in user requests being fulfilled. [Service Operation] [ITIL - Selected Functions]

32. C - While the responsibility for authorization for Changes lies with the Change Manager, they in turn will ensure they have approval from responsible parties in the areas of Finance, Business, and Technology. [Service Transition] [ITIL - Selected Roles]

33. D - A Gradual Recovery, also known as a Cold Recovery, offers >72 hour recovery from a disaster. [Service Design] [ITIL - Generic Concepts and Definitions]

34. A - Mean Time Between Failures is a measure of reliability for repairable products. Mean Time To Failure is a measure of reliability for products that can not be repaired. [Service Design] [ITIL - Technology and Architecture]

35. C - Service-Based SLAs cover a service for all clients. Client-Based SLAs are tailored to the specific needs of a single customer. [Service Design] [ITIL - Selected Processes]

36. B - Short term improvements are those made to the working practices within the Service Operations processes, functions, and the technology itself. Generally they involve smaller improvements that do not change the fundamental nature of a process or technology, such as training, tuning, or personnel redeployments. [Service Operation] [ITIL - Service Lifecycle]

37. C - A Configuration Baseline captures both the structure and details of a configuration and is used as a reference point for later comparison. [Service Transition] [ITIL - Technology and Architecture]

38. D - Key Performance indicators provide insight into the effectiveness and efficiency of the Change Management process. The metrics include number and type of RFCs and Changes implemented, as well as percentage of success or unsuccessful changes. [Service Transition] [ITIL - Selected Functions]

39. D - The benefits of a CMDB include reduction in support costs by using a single system rather than multiple, disparate tools and increased visibility into service offerings, consistency in CI data and access to CI data. [Service Transition] [ITIL - Selected Processes]

40. B - Negotiating and agreeing upon SLAs and OLAs is the responsibility of Service Level Management. {Service Design] [ITIL - Selected Roles]

Knowledge Area Quiz: Generic Concepts and Definitions Practice Questions

Test Name: Knowledge Area Quiz: Generic Concepts and Definitions
Total Questions: 10
Correct Answers Needed to Pass: 7 (70.00%)
Time Allowed: 10 Minutes

Test Description

This practice test specifically targets generic concepts and definitions related to ITIL.

Test Questions

1. Financial Management Partners' IT group has determined that they will not fix a Known Error in an application used by the HR department. Which of the following is a valid reason for not remediating an Known Error?

   A. The implementation team has not finished its analysis.

   B. The root cause has not been found.

   C. The HR division has exceeded its budgeted number of fixes per quarter.

   D. The costs of the fix may exceed the benefits of fixing the error.

2. Connecting Point Software and Diatomic Systems have made an arrangement to host each other's IT infrastructure in the event either of them experiences a disaster that affects their mission critical applications. What is this type of disaster recovery measure called?

   A. Disaster Planning

   B. Gradual Recovery

   C. Quid pro quo

   D. Reciprocal Arrangement

3. Resilience measures what property of a service or component?

   A. The business critical elements of the service

   B. The ability to withstand failure

   C. The ability to function according to requirements

D. The ability to recover from failure

4. The shared services group at Global Storage Technology has developed a service catalog that contains the IT Services, processes, supporting services, and components offered to its customers. This catalog, however, is not intended to be viewed by the customer. What is this catalog called?

A. Technical Service Catalog

B. Operation Catalog

C. Business Service Catalog

D. Configuration Management Catalog

5. Services that provide the basic results or outcome customers require are called:

A. Basic Factors

B. Core Services

C. Value Services

D. Supporting Services

6. How quickly an IT component can be restored to an operational state is known as:

A. Maintainability

B. Serviceability

C. Availability

D. Reliability

7. Bill runs the server operations group at Western Associates. The servers this group maintains are considered production devices and are actively providing services to customers. What is the ITIL term for production components?

A. Live

B. Steady State

C. Live Environment

D. Managed

8. ITIL defines 3 types of change requests. Which of the following is not one of these types?

A. Strategic

B. Normal

C. Standard

D. Emergency

9. John is performing a gap analysis to determine if there is an appropriate balance of security measures for each of the ITIL security perspectives. He is using the Information Security Measure Matrix to perform this assessment. However, he must perform another analysis to identify the level of security required to determine the investment needed in protecting his organization's assets. What is this assessment called?

A. Threat Management

B. Cost-Benefit Analysis

C. Business Impact Analysis

D. Risk Assessment

10. Melissa is documenting availability metrics for her management team. She includes a term that refers to the uptime of a service and calculated as the average time between the recovery of one incident and the occurrence of the next. What is this term?

A. MTRS

B. MTBF

C. MTBR

D. MTBSI

Knowledge Area Quiz: Generic Concepts and Definitions Answer Key and Explanations

1. D - Known Errors have a known underlying cause, and a workaround or permanent solution has been identified. However, it may be possible that the cost of the fix outweighs the benefits of fixing the error. [Service Operation] [ITIL - Generic Concepts and Definitions]

2. D - A reciprocal arrangement is an agreement between two similar sized organizations or businesses to share disaster recovery obligations. [Service Design] [ITIL - Generic Concepts and Definitions]

3. B - Resilience is the ability of a service or component to withstand failure. [Service Design] [ITIL - Generic Concepts and Definitions]

4. A - The Technical Service Catalog contains details of all the IT Services delivered to the customer, together with relationships to the supporting services, shared services, components, and Configuration Items necessary to support the provision of the service to the business. This should underpin the Business Service Catalog and not form part of the customer view. {Service Design] [ITIL - Generic Concepts and Definitions]

5. B - Core services deliver the basic results to the customer. They represent the value that customers require and for which they are willing to pay. [Service Strategy] [ITIL - Generic Concepts and Definitions]

6. A - Maintainability is the measurement of how quickly and effectively a service can be restored to normal functionality and performance after a failure. [Service Design] [ITIL - Generic Concepts and Definitions]

7. A - Live refers to a IT Configuration Item such as a server that is being used to deliver service to customers. A Live Environment is a controlled environment containing Live CI's used to deliver services to customers.[Service Transition] [ITIL - Generic Concepts and Definitions]

8. A - There are three types of change requests: normal, standard, and emergency. Strategic change requests are not defined as a change type by ITIL. [ITIL Generic Concepts and Definitions]

9. B - The cost benefit analysis is used to determine the actual investment in security spending. The cost of reducing or eliminating the threat

should not exceed the value of the asset. [Service Design] [ITIL - Generic Concepts and Definitions]

10. B - MTBF, Mean Time Between Failures, is a measure of uptime, and is an indicator of the reliability of the service. [ITIL - Generic Concepts and Definitions]

ITIL Foundation Practice Exam 11 Practice Questions

Test Name: ITIL Foundation Practice Exam 11
Total Questions: 40
Correct Answers Needed to Pass: 30 (75.00%)
Time Allowed: 60 Minutes

Test Description

This is the eleventh cumulative ITIL Foundation test which can be used as an indicator for overall performance. This practice test includes questions from key ITIL areas.

Test Questions

1. A medium sized brokerage firm has hired an outside firm to review its database logs on a regular basis to determine if proper access processes are being followed. During which phase of the Lifecycle would this type of activity occur?

   A. Continual Service Improvement

   B. Service Operation

   C. Service Strategy

   D. Service Design

2. After a virus outbreak in the main datacenter, the incident response team and senior management of a managed hosting company met to review what went right, what went wrong, and what could be done in the future to prevent occurrences and reduce the amount of time to resolve the problem. What is this review called?

   A. Post-incident Evaluation

   B. Post Mortem

   C. Major Problem Review

   D. Incident Review

3. Service Requirements must be:

   A. PDCA

   B. PPPP

   C. RACI

   D. SMART

4. American Title Company is experiencing an outage on its internet circuit. Their service provider has

indicated that it will not be resolved within the range of the American Title Company's SLA, and as a result the situation is escalated to management for resolution. What is this type of escalation called?

A. Executive Escalation

B. Priority Escalation

C. Functional Escalation

D. Hierarchical Escalation

5. The Service Design team is studying the demand patterns of a service to ensure their internal customers' business plans are synchronized with the organization's service management plans. This type of demand management is called:

A. Technical Demand Management

B. Capacity Management

C. Process Capacity Management

D. Activity-Based Demand Management

6. The Change Advisory Board at Lucas Industries maintains a schedule of approved changes and their proposed implementation dates. What is this schedule called?

A. Projected Service Updates

B. Request for Change

C. Change Log

D. Change Schedule

7. DataMax is releasing a new service to all users in a single operation. What is this release approach called?

A. Automated

B. Big Bang

C. Push Approach

D. Pull Approach

8. The Release Team of Treskett Inc. is considering options for deploying a new service to its users. One method of release is less efficient than the others, but it offers the best options for a completely customized deployment for this release. Which release option is this?

A. Manual

B. Push Approach

C. Phased

D. Pull Approach

9. The Service Asset Manager is responsible for:

A. Full lifecycle management of IT and Service assets from acquisitions to disposal

B. Configuration auditing

C. Recording the relationships between service assets and configuration items

D. Consolidating servers to save money.

10. The Configuration Manager of a state government agency has provided a report on the status of all current and past Configuration Items. What is this activity called?

A. Configuration Status

B. Request for Status

C. Configuration Accounting

D. Status Accounting

11. The Deming Cycle of Continual Improvement involves which steps?

A. Scope, Policies, Reporting, Implementation

B. Plan, Do, Check, Act

C. Design, Pilot, Rollout, Results

D. Validation, Verification, Quality Assurance, Independent Analysis

12. The goals of which process are to manage investments in service management across the enterprise and maximize their value?

A. Service Capacity Management

B. Service Value Management

C. Service Catalog Management

D. Service Portfolio Management

13. Marissa has just joined the IT department at large chain of department stores as part of the Service Design team. Her manager has provided her with a binder that contains all the information needed to provide guidance and structure for the life of a specific service. A new version of this documentation is produced for every new service, major changes to a service, or the retirement of a service. What is this information package called?

A. Service Portfolio

B. Service Catalog

C. Service Index

D. Service Design Package

14. Business plans and strategies, incidents, problems, SLA breaches, and budgets are inputs into what process?

A. Capacity Management

B. Service Catalog Management

C. Service Level Management

D. Continual Service Improvement

15. Eastern Enterprises is working to identify and remediate potential issues with a new service before it is rolled out. In which Lifecycle Phase is this work done?

A. Service Operation

B. Continual Service Improvement

C. Service Design

D. Service Transition

16. Establishing remediation plans in the event of a deployment failure, and creating test plans, happen during which Lifecycle Phase?

A. Service Design

B. Release and Deployment Management

C. Service Operation

D. Service Transition

17. Test plans, deployment plans, schedule, and budget are part of which of the following items?

A. Definitive media library

B. Service Design Package

C. Service Strategy

D. Service Catalog

18. Which process assures that all goals for a specific service are agreed upon by all responsible parties?

A. Service Design Management

B. Service Level Management

C. Service Value Management

D. Capacity Management

19. The IT division of a multinational accounting firm has entered an agreement with the UK operations division to provide specific IT services and support. What is this internal agreement called?

A. Service Level Requirements

B. Operational Level Agreement

C. Service Level Agreement

D. Underpinning contract

20. Jennifer is reviewing a document that contains only the services that are in currently in operation. What is this document called?

A. Service Catalog

B. Service Portfolio

C. Service Menu

D. Service Listing

21. Lifeline Industries has developed a matrix to define the roles and responsibilities of individual staff or entire groups in relationship to processes and activities. What is the matrix called?

A. RACI Model

B. Organization Chart

C. Service Matrix

D. Rummler-Brache Diagram

22. Service Level Management is found within which two Lifecycle Phases?

A. Service Operation and Service Design

B. Service Transition and Service Design

C. Service Design and Continual Service Improvement

D. Service Strategy and Service Design

23. Service Value is defined as the combination of which two concepts?

A. Service Performance and Service Capabilities

B. Service Packages and Service Definitions

C. Service Use and Service Design

D. Service Warranty and Service Utility

24. Which of the following metrics is calculated as a percentage of agreed upon service time minus downtime?

A. Availability

B. Performance

C. Reliability

D. Capacity

25. Nora is responsible for recommending improvements to a service. What role is she filling?

A. Service Strategist

B. Service Manager

C. Service Owner

D. Service Designer

26. Service Level Agreement negotiation occurs in which Lifecycle Phase?

A. Continual Service Improvement

B. Service Transition

C. Service Operation

D. Service Design

27. Enrique is reviewing the version number of a Configuration Item in the CMDB. The version number is an:

A. Auditable record

B. Assembly

C. Attribute

D. Asset

28. During which Lifecycle Phase are supplier contracts renewed or terminated?

A. Service Transition

B. Continual Service Improvement

C. Service Operation

D. Service Design

29. In what Lifecycle Phase are the majority of Supplier Management process activities performed?

A. Service Design

B. Service Operation

C. Service Transition

D. Continual Service Improvement

30. Eclipse Systems offers reassurance to its customers that it will meet agreed upon requirements for availability, capacity, continuity, and security. What is this reassurance called?

A. SLA

B. Warranty

C. Guarantee

D. Rating

31. Categorization of an incident is based on which two factors?

A. Outage Duration and Number of Users Affected

B. Impact and Urgency

C. SLA and Guarantees

D. Value and Warranty

32. The intangible assets that MaxStorage uses to manage its IT services are known as capabilities. How are these capabilities acquired?

A. Capabilities Management Processes

B. They are developed and matured over time

C. Standard procurement processes

D. They are outsourced

33. License management, baseline configurations, and application version management are handled by which Configuration Management activity?

A. Verification and Audit

B. Control

C. Status accounting

D. Identification

34. Eastern Enterprises IT division has created a catalog containing the details of the IT services delivered to their customers. This catalog defines the relationships between Eastern Enterprise's business divisions and the processes that are supported by an underlying IT service. What is this catalog called?

A. Support Service Catalog

B. Technical Service Catalog

C. Business Service Catalog

D. Division Service Catalog

35. The law firm of Tregoe and Lawton has a large number of case records to be entered into their document management system on a daily basis. The amount of bandwidth used to send imaged documents to the documentation repository slows down all other functions during peak use times. The IT department determines the most cost effective way to eliminate this problem is to batch the images for transfer after hours. What is this called?

A. Task Switching

B. Job Sharing

C. Batch Processing

D. Job Scheduling

36. An escalation based on knowledge or skills is referred to as what?

A. Tier 2 Escalation

B. Engineering Escalation

C. Hierarchical Escalation

D. Functional Escalation

37. Which of the following are risks associated with Service Catalog Management?

A. An inaccurate Service Catalog

B. An accurate Service Catalog

C. Users are familiar with the services delivered

D. Users are not familiar with the services delivered

38. Which of the following indicates that service is fit for the purpose for which it was designed?

A. Value

B. Available

C. Utility

D. Warranty

39. The Service Strategy team are reviewing the requirements for availability, capacity, continuity, and security for a potential new service offering. What key service element do these properties create?

A. Value

B. Utility

C. Design

D. Warranty

40. Washburn Electronics has implemented a number of databases, documentation repositories, configuration management systems, and reporting tools to used to store, view, manage, and use information about its IT services. What are these tools and systems collectively called?

A. Configuration management system

B. Service knowledge management system

C. Definitive media library

D. Service design package

ITIL Foundation Practice Exam 11 Answer Key and Explanations

1. A - There are three types of change requests: normal, standard, and emergency. Strategic change requests are not defined as a change type by ITIL. [The Service Lifecycle]

2. C - After every major problem, while memories are still fresh, a review should be conducted to capture lessons learned. [Service Operation] [ITIL - Selected Processes]

3. D - Service Requirements must be SMART: Specific, Measurable, Achievable/ Appropriate, Realistic/ Relevant, and Timely/ Timebound. [Continual Service Improvement] [ITIL - Generic Concepts and Definitions]

4. D - Hierarchical or Vertical Escalations are used to escalate an issue to authorized line management when resolution of an incident will not be in time or satisfactory according to the terms of the SLA. [Service Operation] [ITIL - Selected Processes]

5. D - Activity-based demand management is the synchronization of customer business process demand with the service provider's management plans, to ensure that the supply of services matches customer demand. [Service Strategy] [ITIL - Selected Processes]

6. D - The Change Schedule is a schedule of approved changes and proposed implementation dates. [Service Transition] [ITIL - Selected Roles]

7. B - The deployment of a new or changed service to all users in one single operation is called the Big Bang approach. [Service Transition] [ITIL - Key Principles and Models]

8. A - Manual Releases are more likely to be inefficient and error prone than other releases, due to the impact of repeated manual activities. [Service Transition] [ITIL - Selected Processes]

9. A - Service Asset Management is responsible for the management of service assets across the whole lifecycle and maintenance of the asset inventory.[Service Transition] [ITIL - Selected Roles]

10. D - Reporting of all current and historical data for each Configuration Item through its lifecycle is known as Status Accounting. [Service Transition] [ITIL - Selected Functions]

11. B - The ITIL CSI Improvement Process is based on Deming's Cycle of Continual Improvement: Plan, Do, Check, Act. [Continual Service Improvement] [ITIL - Generic Concepts and Definitions]

12. D - The goals of Service Portfolio Management are to realize and create maximum value, while minimizing risks and costs.[Service Strategy] [ITIL - Selected Processes]

13. D - The information contained within a Service Design Package includes all aspects of the service and its requirements, and is used to provide guidance through all the subsequent stages of its lifecycle. It contains information such as requirements, user acceptance criteria, and service transition plans. [Service Design] [ITIL - Key Principles and Models]

14. A - Capacity Management is critical for ensuring the effective and efficient capacity and performance of services in line with business requirements and overall IT strategic objectives. [Service Design] [ITIL - Selected Processes]

15. D - The Service Transition phase is the bridge between Service Design and Service Operation. Functional and technical errors not found in this phase can significantly impact the business or the infrastructure, and typically cost more to remediate in operation than in pre-production. [Service Transition] [ITIL - Service Lifecycle]

16. D - Service Transition is responsible for the planning of fail situations and the test plans that are needed to ensure a service is running appropriately before it is deployed. [Service Transition] [ITIL - Service Lifecycle]

17. B - A Service Design Package (SDP) contains the requirements and relevant design/ build/ deploy/ support information for a new service. A new SDP is created for every new service, as well as when a service has major changes or is retired. [ITIL - Service Lifecycle]

18. B - Service Level Management assures reliable communication with all responsible parties and maintains the relationships with those parties. It agrees upon the goals of the service provision of these parties and provides the management information required to attain those goals. [Service Design] [ITIL - Selected Processes]

19. B - Operational Level Agreements (OLAs) are internal agreements which support the IT organization in their delivery of services. [Service Design]

[ITIL - Service Management as a Practice]

20. A - The Service Catalog is a subset of the Service Portfolio and consists of only active services in operation. The Service Portfolio represents all active and inactive services in the various phases of the lifecycle. [Service Design] [ITIL - Selected Processes]

21. A - The RACI Model is used to define roles and responsibilities of people or groups in relation to processes and activities. It defines who is responsible, accountable, consulted, and informed for a specific action or event.[Service Design] [ITIL - Key Principles and Models]

22. C - Service Level Management is a process that is found within two Service Lifecycle phases: Service Design and Continual Service Improvement. [Service Design] [ITIL - Selected Processes]

23. D - Service Value is the combination of Service Warranty and Service Utility [Service Strategy] [ITIL - Generic Concepts and Definitions]

24. A - Availability is the ability of a service to perform its intended function when required, and is based on its reliability, maintainability, serviceability, and security. [The Service Lifecycle]

25. C - The Service Owner bears the responsibility for the initiation, transition, and maintenance of a service, and identifies improvement points for the service he or she owns. [Service Strategy] [ITIL - Selected Roles]

26. D - Within Service Design, Service Level Management is concerned with negotiating and agreeing upon SLAs. [Service Design] [ITIL - Service Lifecycle]

27. C - An Attribute is a piece of information about a Configuration Item. Examples are name, location, version number, and cost. Attributes of CIs are recorded in the Configuration Management Database. [Service Transition] [ITIL - Generic Concepts and Definitions]

28. C - Supplier Contracts are renewed or terminated during Service Operation. [Service Operation] [ITIL - Service Lifecycle]

29. A - Supplier management is strongly rooted within the Service Design phase, but some process activities occur in other Lifecycle Phases. [Service Design] [ITIL - Selected Processes]

30. B - Service Warranty provides the customer a level of reassurance and guarantee to meet agreed requirements. [Service Strategy] [ITIL - Service Management as a Practice]

31. B - Categorization is the statistical aspect of prioritization. It is based on the impact (degree to which the user is affected) and urgency (degree to which resolution can be delayed). [Service Operation] [ITIL - Selected Processes]

32. B - Capabilities of an organization to manage services are intangible assets that cannot be purchased, but instead must be developed and matured over time. [Service Design] [ITIL - Key Principles and Models]

33. B - Configuration Control ensures that CIs are adequately managed. No CI can be added, changed, replaced, or removed without following the proper procedure. [Service Transition] [ITIL - Selected Functions]

34. C - The Business Service Catalog contains details of all the IT services delivered to the customer, together with the relationships to the business units and the business processes that rely on the IT services. This is the customer view of the Service Catalog.[Service Design] [ITIL - Selected Processes]

35. D - Job scheduling is the planning and management of software tasks that are required as part of an IT Service. Job Scheduling is carried out by IT Operations Management and is often automated using software tools that run batch or online tasks at specific times of the day, week, month, or year. [Service Operation] [ITIL - Technology and Architecture]

36. D - Functional Escalations are based on knowledge or expertise, and are also known as Horizontal Escalations.[Service Operation] [ITIL - Generic Concepts and Definitions]

37. A - Risks associated with Service Catalog Management include inaccurate information in the Service Catalog, acceptance of the Service Catalog and its use in operations processes, and accuracy of information supplied by the business, IT, and Service Portfolios. [Service Design] [ITIL - Selected Processes]

38. C - Service Utility defines the functionality of an IT Service from the customer's perspective, ensuring that the desired performance is supported and constraints have been removed. [Service Strategy] [ITIL - Key Principles and Models]

39. D - Warranty guarantees the utility of a service by ensuring that it is available and offers sufficient capacity, continuity, and security. [Service Strategy] [ITIL - Key Principles and Models]

40. B - The service knowledge management system (SKMS) is the set of tools, databases, configuration management systems, and other information management systems used to collect, store, view, manage, and update information needed to manage all phases of an IT service's lifecycle. [ITIL - Service Lifecycle]

Knowledge Area Quiz: Selected Processes Practice Questions

Test Name: Knowledge Area Quiz: Selected Processes
Total Questions: 20
Correct Answers Needed to Pass: 14 (70.00%)
Time Allowed: 20 Minutes

Test Description

This extended practice test targets ITIL concepts related to Selected Processes.

Test Questions

1. A pre-approved change that is low risk and follows a set of procedures, such as creating a new user account, is what type of change?

    A. Request for Change

    B. Service Request

    C. Normal Change

    D. Standard Change

2. The senior IT team of a manufacturing company is reviewing its service portfolio in an effort to revamp their service offerings. Which of the following is not a valid outcome for an existing service?

    A. Rationalize

    B. Refactor

    C. Retool

    D. Retire

3. Which service operation process ensures the elimination of the root cause of a problem causing an outage?

    A. Event Management

    B. SLA

    C. Problem Management

    D. Incident Management

4. The senior management of West Wind Imports is trying to balance the need for stability and availability with responsiveness. Which Lifecycle Phase deals with this conflict?

    A. Service Transition

B. Service Design

C. Service Operation

D. Service Strategy

5. Which of the following is an example of a Configuration Item?

A. Server configs

B. RFCs

C. SLAs

D. All of the above

6. Analyzing and tracking patterns of business activities make it possible to do what?

A. Restrict non-critical activities during peak periods

B. Stagger work start times

C. Prioritize reports and batch jobs

D. Predict demand for services that support the process

7. An organization has outsourced its network design and deployment services to a large integrator. The lessons learned from past design and deployment efforts have led to the development of a plan to improve and enhance the service being offered. Where are these plans recorded?

A. Supplier Performance Reports

B. Supplier Quality Reports

C. Business Process Outsourcing

D. Supplier Service Improvement Plans

8. Systemix provides managed hosting services. They have a basic hosting package that includes a shared server, 500GB of storage, and domain name registration. Additional services, including static IPs and reporting tools, are available. What are these additional services called?

A. Service Capabilities

B. Shared Services

C. Supporting Services

D. Service Enhancements

9. Data Storage Systems has three categories of services defined in its Service Portfolio. These categories refer to proposed services, current services, and services that are no longer available. To which ITIL service categories do these map?

A. Gold, Silver, Bronze

B. Service Pipeline, Service Catalog, Retired Services

C. Pre-implementation Services, Implementation Services, Post-implementation Services.

D. Tier 1, Tier 2, Tier 3

10. The management of ongoing service performance as detailed in Service Level Agreements occurs in which process?

A. Business Capacity Management

B. Service Capacity Management

C. Ongoing Capacity Management

D. Component Capacity Management

11. What process is responsible for authorizing users or preventing users from using a Service?

A. Verification

B. Access Management

C. Rights Management

D. Auditing

12. Dynamic Data implemented a change in its network, but discovered that the change caused unplanned service interruptions. Because the extent of the interruptions is unknown, the implementation team decides to abort the change and recover to the last known good configuration noted in the change plan. What is this process called?

A. Remediation

B. Outcome Facilitation

C. Service Incident

D. Modeling

13. The security team within a manufacturing corporation has discovered that their web server has been taken offline by a Denial of Service attack. They immediately take the server off the network and conduct an analysis of the source and cause of the attack, followed by a rebuild of the server. What is this Security Management process called?

A. Detection

B. Correction/ Recovery

C. Threat Management

D. Control

14. Problem Management is composed of two sub processes. What are these sub processes called?

A. Standard and Emergency Problem Management

B. Major and Minor Problem Management

C. Incident Management and Event Management

D. Proactive and Reactive Problem Management

15. Identification of critical business processes, potential damage or loss from disruption, resources required to continue critical business processes, maximum permissible total downtime, and maximum permissible time till complete recovery are steps in which Service Continuity Management activity?

A. Business Impact Analysis

B. Risk Reduction

C. Business Continuity and Disaster Recovery

D. Risk Assessment

16. Controlling demand for a service can be done which two ways?

A. Internal Constraints, External Constraints

B. Technical Constraints, Financial Constraints

C. Calculation-based Constraints, Impact-based Constraints

D. Strategic Constraints, Tactical Constraints

17. What are the three strategic categories a service investment could fall into?

A. Cost Savings, Cost Avoidance, Balance Spending

B. Business Process Alignment, Risk Reduction, Business Transformation

C. Service Improvement, New Service, Service Retirement

D. Transform the Business, Grow the Business, Run the Business

18. ITIL is concerned with which three types of SLAs?

A. Promotional, Permanent, Interim

B. Service-based, Customer-based, and Multilevel

C. Corporate-level, Customer-level, and Service-level

D. Transaction-level, Reporting-level, Availability-level

19. Western Enterprises is evaluating new technology for implementation in their service environment. Which of the following processes is concerned with this evaluation?

A. Technical Capacity Management

B. Service Capacity Management

C. Component Capacity Management

D. Engineering Capacity Management

20. The four perspectives of Security Management are:

A. Accounting, HR, Legal, IT

B. Organizational, Procedural, Physical, Technical

C. Network, Server, Backups, Physical

D. Business, Financial, Physical, IT

Knowledge Area Quiz:
Selected Processes
Answer Key and Explanations

1. D - A pre-approved change is low risk, relatively common, and follows a procedure or work instruction. [Service Transition] [ITIL - Selected Processes]

2. C - The outcomes of a service portfolio review of existing services are renew, replace, retain, refactor, retire, and rationalize. [Service Strategy} [ITIL - Selected Processes]

3. C - The goal of problem management is to minimize the adverse impact of problems on the business that are caused by errors within the IT infrastructure and to prevent the recurrence of incidents related to these errors. [Service Operation] [ITIL - Selected Processes]

4. C - Service Operation balances the need to maintain the status quo against the need to continuously adapt to change. [Service Operation] [ITIL - Selected Processes]

5. D - A Configuration Item is any component that supports an IT service, including IT components, RFCs, Incident Records, and SLAs. [Service Transition] [ITIL - Selected Processes]

6. D - Business processes are the primary source of demand for services. Patterns of business activity influence the demand patterns seen by the service providers. The study of these patterns is important for effective capacity management. [Service Strategy] [ITIL - Selected Processes]

7. D - Supplier Service Improvement Plans are used to record all actions and plans agreed between suppliers and service providers. [Service Design] [ITIL - Selected Processes]

8. C - A Service Package is made up of the core services provided, additional supporting services that are available, and the service levels. This modular approach allows service providers to avoid one-size-fits all offerings, while still standardizing the services offered. [Service Strategy] [ITIL - Selected Processes]

9. B - The three categories of services defined in the Service Portfolio are: Service Pipeline, Service Catalog, and Retired Services. [Service Strategy] [ITIL - Selected Processes]

10. B - Service Capacity Management focuses on managing ongoing service performance as detailed in the SLA,

and establishes baselines and profiles of service usage. [Service Design] [ITIL - Selected Processes]

11. B - Access management is concerned with granting authorized users the right to use a Service, while simultaneously preventing access to non-authorized users, thus protecting the confidentiality, integrity, and availability of information and infrastructure. [Service Operation] [ITIL - Selected Processes]

12. A - Remediation is recovery to a known state after a failed Change or Release. [Service Transition] [ITIL - Selected Processes]

13. D - Information Security is enforced in an organization by the Control Process. This process consists of the following steps: Prevention/ Reduction, Detection/ Repressing/ Correction/ Recovery, and Evaluation. [Service Design] [ITIL - Selected Processes]

14. D - The Activities of Problem Management are carried out within the Proactive and Reactive Problem Management processes. The main goal of Proactive Problem Management is to identify errors through trend analysis that might otherwise be missed, and then take preventative action. Reactive Problem Management is similar, but occurs after a problem is discovered. Subsequent reactive activities include root cause analysis and Known Error correction. [Service Operation] [ITIL - Selected Processes]

15. A - Business Impact Analysis is concerned with the identification of critical business functions, the consequences of the loss of those functions, and the recovery requirements for those functions. [Service Design] [ITIL - Selected Processes]

16. B - Demand for a service can be influenced or managed through technical constraints such as bandwidth throttling or financial constraints such as charging higher service rates for usage during peak hours.[Service Strategy] [ITIL - Selected Processes]

17. D - Service investments are split into three strategic categories: Transform the Business, Grow the Business, and Run the Business. [Service Strategy] [ITIL - Selected Processes]

18. B - Three types of SLA structures discussed within ITIL are Service-based, Customer-based, and Multilevel. [Service Design] [ITIL - Selected Processes]

19. C - Component Capacity Management identifies and manages each of the components of the IT Infrastructure such as CPU, memory, bandwidth and evaluates new technologies. [Service Design] [ITIL - Selected Processes]

20. B - Security Management must consider the following four perspectives: Organizational, Procedural, Physical, and Technical. [Service Design] [ITIL - Selected Processes

ITIL Foundation Practice Exam 12 Practice Questions

Test Name:
ITIL Foundation Practice Exam 12
Total Questions: 40
Correct Answers Needed to Pass:
30 (75.00%)
Time Allowed: 60 Minutes

Test Description

This is the eleventh cumulative ITIL Foundation test which can be used as an indicator for overall performance. This practice test includes questions from key ITIL areas.

Test Questions

1. Which of the following CORRECTLY describe the 4 Ps of ITSM?

   A. Profession-Planning-Process-Pride

   B. Partners-Planning-Products-Process

   C. Partners-People-Products-Process

   D. Partners-People-Profession-Process

2. Which of the following are the CORRECT choices for the stages in Deming Quality Cycle?

   A. Process-Do-Check-Audit

   B. Plan-Do-Check-Act

   C. Produce-Data-Correct-Audit

   D. Plan-Data-Check-Act

3. What is a common method for prioritizing an Incident in ITIL?

   A. Prioritize the Incident based on the Impact and Urgency of the Incident.

   B. Prioritize the Incident based on First-Come-First-Serve basis.

   C. Allow the Incident to follow the normal route of action and escalation and closure.

   D. Prioritize the Incident based on the person's position within the organization who raised the Incident.

4. RACI model is a useful tool which helps in which of the following activities?

   A. RACI model is used in designing activities and functions.

   B. RACI model is used in designing an organization's policies.

   C. RACI model is used in designing processes.

   D. RACI model is used in only measuring the effectiveness of processes.

5. Security failures are caused by which of the following?
   I. Technical errors
   II. Human errors
   III. Procedural errors
   IV. Database failures

   A. I, II and III

   B. II, III and IV

   C. I, III and IV

   D. I, II and IV

6. THE R-A-C-I in the RACI model stand for which one of the following?

   A. RACI stands for Responsibility-Accounts-Consult-Inform.

   B. RACI stands for Responsibility-Accountability-Consult-Information.

   C. RACI stands for Responsibility-Accountability-Consult-Inform.

   D. RACI stands for Reform-Accountability-Consult-Inform.

7. What terms are represented by the "7 Rs" of the Change Management process?

   A. Raised, Return, Run, Risks, Required, Responsible, Relationship

   B. Raised, Reason, Return, Risks, Review, Responsible, Relationship

   C. Raised, Reason, Return, Risks, Required, Responsible, Relationship

   D. Raised, Reason, Return, Risks, Reduce, Responsible, Relationship

8. Which of the following is NOT an activity of the Service Design life cycle stage?

A. Production and maintenance of IT policies and design documents, including designs, plans, architectures and policies.

B. Requirements collection, analysis and engineering to ensure that business requirements are clearly

C. Review and revision of all processes and documents involved in Service Design, including designs, plans, architectures and policies.

D. Enabling the organization to influence the management decision making by ensuring the availability of reliable and secure information throughout the Service life cycle.

9. What is the purpose of SFA(Service failure Analysis) in the Service Design stage of the Service Life Cycle?

I. Identification of the service interruptions

II. Resolution of the service Interruptions

III. Identification of the service breakdowns

IV. Evaluation of the service intervals

A. I only

B. II only

C. III only

D. None of these statements reflect the purpose of the SFA in Service Design

10. Who is the originator of the "Plan-Do-Check-Act Quality Cycle" for Service Improvements?

A. Edward Deming

B. Jeroen Bronkhorst

C. Kaizen

D. Robert Ishikawa

11. What type of SLA would contain "corporate level", "custom level", and "service level" sub types?

A. Customer Based SLA

B. Multi-level Based SLA

C. Supplier-based SLA

D. Service based SLA

12. Which of the following choices correctly classify the CIs?

A. Service CIs, Service Life Cycle CIs, Organizational CIs, Internal CIs, External CIs, Interface CIs.

B. Service CIs, Service Life Cycle CIs, Internal CIs, External CIs, Interface CIs.

C. Service CIs, Procedural CIs, Organizational CIs, Internal CIs, External CIs, Interface CIs.

D. Service CIs, Process CIs, Organizational CIs, Internal CIs, External CIs, Interface CIs.

13. The Service Knowledge Management system is introduced in which phase of the Service Life Cycle?

A. Service Strategy

B. Service Design

C. Service Operation

D. Service Transition

14. Which of the following is not a 'Function' as defined by ITIL?

A. Service Desk

B. Access Management

C. Application Management

D. Technical management

15. To whom does a Service in ITIL provide value?

- Customer
- Service provider
- IT
- General public

A. Customer and Service Provider

B. Service Provider and IT

C. Customer, Service Provider, IT and general public

D. Customer, Service Provider and IT

16. In which stage of the Service Life cycle does the customer see the quality of the service come to life in everyday use of the services?

A. Service Operation

B. Service design

C. Service strategy

D. Service Transition

17. Which of the following statements are TRUE with respect to the Change Management Process in the Service Transition phase?

I. Respond to business and customer change requests on a fixed date and time.
II. Implement changes as per the agreed SLAs in a cost effective manner.
III. Comply to governance, legal, contractual and regulatory needs.
IV. Try to reduce the total number of failed changes by effectively implementing them and therefore reducing service disruption.

A. I, III and IV

B. All of these statements are true

C. II, III and IV

D. I and III

18. Which of the following is the CORRECT order of activities in the Service Request Fulfillment process?

I. Initially, the service request is checked for appropriate approval and then after categorization, it is prioritized.
II. The Service request is then recorded as per the procedures.
III. The request is worked upon and tracked for completion. If the requestor is satisfied with the action, then it is closed.
IV. The Service request is then handed over to the department owners as per existing procedures.

A. II-I-IV-III

B. II-I-III-IV

C. I-II-IV-III

D. I-II-III-IV

19. Which stage of the Service Life cycle is viewed as the 'factory' of IT?

A. Service Operation

B. Service Transition

C. Service strategy

D. Service design

20. The supporting activities like management of the infrastructure, datacenters, vendors, databases, staff, training, up-skilling etc are listed under which phase of the Service Life Cycle?

A. Service Operation

B. Service Design

C. Service Strategy

D. Service Transition

21. Which of the following are the characteristics of a best partnership achieved in a Service Provider and Customer relationship?

A. Evolve and grow together.

B. Share the risk, reward as well as evolve and grow together.

C. Share the reward however small or big it is.

D. Share the risk with the customer.

22. Which of the following statement/s are TRUE with respect to Configuration items?

I. CIs may be grouped and managed together.

II. CIs should be selected, classified and identified so that they are traceable throughout the Service Life cycle.

III. CIs may vary largely in complexity, size and type.

IV. CIs can be a piece of hardware or a module of software.

A. II, III and IV

B. All of them

C. I, III and IV

D. I, II and III

23. What are the two types of monitoring tools available for use in the Event management Process?

A. Software and Hardware

B. Normal and automated

C. Active and Passive

D. Primary and Secondary

24. Operations Control and Facilities Management form types of which function within the Service Operation phase of the Service Lifecycle?

A. Applications

B. IT Operations management

C. IT Service desk

D. Technical Management

25. Which one of the following statements do NOT form an output from Service V model of the Service Validation and Testing Process?

A. Test results and Analysis of test results.

B. Test models and test activity details.

C. Testing environment details and the configuration baselines.

D. Test plan and design.

26. To work out the price for a service, which of the following things must be taken into consideration?

A. All of them

B. The available market must be considered too.

C. The relevance of the service to the customers must be considered.

D. The existing competition for the service must be considered.

27. Which of the following statement/s are TRUE with respect to the IT Service Industry?

I. Over a period of time all services, including the newly designed ones, undergo changes and improvements.

II. As business and business needs change, business outcomes also change.

III. To improve the service and service quality, investment is always sought after in new technologies

IV. Information should always be recorded as events, incidents or problems

A. II, III and IV

B. I, II and IV

C. I, II and III

D. None of these statements are true

28. Which of the following statement/s are TRUE with respect to an event in the Event Management Process?

I. The occurrence of an event has considerable significance on the infrastructure and IT service delivery.

II. Events are notifications created by a CI (Configuration

Item) or IT service or a software tool.

III. Events can be monitored using active and passive monitoring tools.

A. II, III

B. II only

C. I, III

D. I, II

29. Which of the following statements do NOT form an input to the 7 Step Improvement Process?

A. The desire to improve operational efficiency

B. The need to improve

C. The drive to reduce cost

D. The need to do business

30. Which of the following statements is FALSE when deciding the appropriate level for release unit within an organization?

A. The level of the release unit depends on the storage available in test and production environment.

B. The level of the release unit depends on the size of the organization.

C. The level of the release unit depends on the time need to implement the change and the resource needed to execute it.

D. The level of the release unit depends on the ease and amount of change.

31. Information Security Management process is successfully implemented in an organization when which the following statements are TRUE?

I. Information is available as and when needed

II. Information is secured with appropriate level of confidentiality

III. Information is complete, accurate and is protected from unauthorized access

IV. Information is exchanged between organization and others in a trusted, authenticated and reliable way

A. II, III and IV

B. I, II and IV

C. I, II and III

D. All of the statements are true

32. Which of the following are the goals of the Transition Planning and Support process in Service Transition phase of the Service Life cycle?

I. To ensure effective realization of Service requirements from Service Strategy and Service Design stages.

II. Throughout the Service Transition activities, failure risks and disruptions are identified and managed.

A. I only

B. II only

C. I and II

D. None of these statements reflect the goals of Transition Planning and Support

33. Which of the following improvements offers a short-term benefit to an organization?

A. Changes to the Internet access for all employees

B. A two-day IT training course for employees

C. New word processing software for all desktops

D. Server upgrades

34. Which of the following is a primary goal of the Access Management process as described in ITIL?

A. The primary goal of Access management process is to prevent all users from accessing the admin login of the software and database.

B. The primary goal of Access management process is to protect the Security of the information assets and the IT infrastructure.

C. The primary goal of Access management process is to not allow the employees to take any print outs or data storage media to and from the organization without prior permission.

D. The primary goal of Access management process is to allow access only to employees and to restrict access to anybody including visitors within an organization.

35. Which of the following statement/s are FALSE with respect to the IT Service Industry?

I. Whenever a new service is conceived and launched, all the activities ensure that business value is retained and it produces results as expected
II. When a new service is introduced, there are always ways to look at it in the form of improvement opportunities.
III. Continual Service Improvement process feeds back only to Service Strategy stage in the Service Lifecycle.

A. II

B. III

C. All of these statements are false

D. I

36. Which one of the following statements is the primary goal of the Service Operation?

I. To create and manage services
II. To deliver and support services
III. To test and support services
IV. To deliver and create services

A. I

B. III

C. II

D. IV

37. Who issues the final decision with respect to the change during a Change Advisory Board Meeting?

A. The Change Manager issues the final decision with respect to the change.

B. The Test Manager issues the final decision with respect to the change.

C. The Manager who raised the change issues the final decision with respect to the change.

D. The Release Manager issues the final decision with respect to the change.

38. What is the objective of the Service Catalog Management process?

A. documented and agreed.

B. To capture metrics that will enable better decision-making and continual improvement of service management practices in the design stage of the Service Lifecycle.

C. To capture the design of the measurement methods and metrics of the services, as well as the architectures and its processes.

D. To manage the information within the Service Catalog and to ensure its correctness for implementation in the LIVE environment.

39. What is the main purpose of the Service Design Stage in the Service Life Cycle?

A. To design a new or updated service which will be introduced into the LIVE environment.

B. To design the outputs to the next stage of the life cycle.

C. To see that the inputs from the previous stage are arrived properly.

D. To design a service solution for the customer.

40. Service or Business value is always in the perception of the customer. This statements translates to and leads us to which of the following statements?

A. If you give (good or bad) service to the customer, he will give you more business.

B. If the services generate more value to the customer, this strengthens the business relations and the bond between the business and the customer.

C. You must always please the clients to get more business.

D. If the services are available at very cheap cost, the customer will be too pleased to give repeat orders and recommend more clients.

ITIL Foundation Practice Exam 12 Answer Key and Explanations

1. C - Partners-People-Products-Process correctly describe the IV)Ps of ITSM. [Service Management as a Practice]

2. B - The CORRECT choices for the stages in Deming Quality Cycle are - Plan-Do-Check-Act. (Continual Service Improvement) - [ITIL Key Principles and Models]

3. A - In ITIL, an Incident is often prioritized based on the Impact and Urgency of the Incident. This is called prioritizing an Incident. (Service Operation) - [ITIL Selected Functions]

4. C - RACI model is used in designing processes. (ITIL Generic Concepts) - [ITIL Generic Concepts and Definitions]

5. A - Security failures are caused by Technical errors, Human errors and Procedural errors. (Service Design) [ITIL Selected Processes]

6. C - RACI stands for Responsibility-Accountability-Consult-Inform. (ITIL Generic Concepts) [ITIL Generic Concepts and Definitions]

7. C - The 7 Rs of the Change Management process are Raised, Reason, Return, Risks, Required, Responsible, Relationship. (Service Transition) [ITIL Selected Processes]

8. D - The following are the activities of the Service Design life cycle stage. I).Requirements collection, analysis and engineering to ensure that business requirements are clearly documented and agreed. II).Review and revision of all processes and documents involved in Service Design, including designs, plans, architectures and policies. III). Production and maintenance of IT policies and design documents, including designs, plans, architectures and policies. (Service Design) – [Service Lifecycle]

9. A - SFA (Service failure Analysis) in Service Design helps in identifying the cause of service interruptions. (Service Design) [ITIL Selected Processes]

10. A - Edward Deming is the originator of the "Plan-Do-Check-Act Quality Cycle". Otherwise known as the Deming Cycle. (Continual Service Improvement) [ITIL Generic Concepts and Definitions]

11. B - Multilevel based SLA is classified as Corporate, Customer and Service level SLAs. (Service Design) [ITIL Selected Processes]

12. A - These choices correctly classify the CIs - Service CIs, Service Life Cycle CIs, Organizational CIs, Internal CIs, External CIs, Interface CIs. (Service Transition) [ITIL Selected Processes]

13. D - The Service Knowledge Management system is introduced in Service Transition phase of the Service Life Cycle. (Service Transition) [ITIL Generic Concepts and Definitions]

14. B - Access management is considered a "process" as per ITIL), not a function. (ITIL Generic Concepts) [ITIL Selected Functions]

15. D - Service in ITIL provides value to Customer, Service Provider and IT (ITIL Generic concepts) [ITIL Generic Concepts and Definitions]

16. A - Service Operation stage of the Service Life cycle is where the customer see the quality of the service come to life in everyday use of the services. (Service Operation) [ITIL Selected Processes]

17. C - The following statements are TRUE with respect to the Change Management Process in the Service Transition phase.
I). Respond to business and customer change requests on a priority basis. II). Implement changes as per the agreed SLAs in a cost effective manner. III). Comply to governance, legal, contractual and regulatory needs. IV). Try to reduce the total number of failed changes by effectively implementing them and therefore reducing service disruption. (Service Transition) [ITIL Selected Processes]

18. A - The following is the CORRECT order of activities in the Service Request Fulfillment process. (Service Operation)
I). The Service request is recorded by the Service Desk personnel. II). The service request is checked for appropriate approval and is categorized and prioritized. III). The Service request is handed over to respective department/owners and auctioned as per existing procedures. It may be escalated if not resolved quickly. IV). The request is tracked for completion and closed after the requestor is satisfied with the action. [ITIL Selected Processes]

19. A - Service operation stage if viewed as the 'factory' of IT. (Service Operation) [ITIL Selected Processes]

20. A - The supporting activities like management of the infrastructure, datacenters, vendors, databases, staff, training, up-skilling etc are listed under Service Operation phase of the Service Life Cycle. (Service Operation) [ITIL Selected Processes]

21. B - The best partnership achieved in a Service Provider and Customer relationship indeed has the characteristics of sharing the risk and reward as well as evolving and growing together. [ITIL Generic Concepts and Definitions]

22. B - All the following statement/s are TRUE with respect to Configuration items. I). CIs may be grouped and managed together. II). CIs should be selected, classified and identified so that they are traceable throughout the Service Life cycle. III). CIs may vary largely in complexity, size and type. IV). CIs can be a piece of hardware or a module of software. (Service Transition) [ITIL Selected Processes]

23. C - Active and Passive are the two types of monitoring tools available for use in the Event Management Process. (Service Operation) - Selected Processes [ITIL Selected Processes]

24. B - Operations Control and Facilities Management form types of IT Operations Management function within the Service Operation phase of the Service Lifecycle. (Service Operation) - Selected Functions [ITIL Selected Functions]

25. D - Test plan and design is not an output from the Service Validation and testing process. (Service Transition) - Selected Processes [ITIL Selected Processes]

26. A - To work out the price for a service, all the following things must be taken into consideration - The relevance to the customer, available market and the competition. (ITIL Service management as a practice) [Generic Concepts and Definitions]

27. C - The following statements are TRUE with respect to the IT Service Industry. I). Over a period of time all services including the newly designed ones undergo changes and improvements. II). As business and business needs change, business outcomes also change. III). To improve the service and service quality, investment is always sought after in new technologies [ITIL Service Management as a practice]

28. D - The following statements are true with respect to an event in the Event Management Process.
I). The occurrence of an event has considerable significance on the infrastructure and IT service delivery.

II). Events are notifications created by a CI (Configuration Item) or IT service or a software tool. (Service Operation) [ITIL Selected Processes]

29. D - The need to do business is NOT an input to the 7 Step Improvement Process. (Continual Service Improvement) [ITIL Selected Processes]

30. B - The size of the organization does not matter when deciding the appropriate level for release unit within an organization. (Service Transition) [ITIL Selected Processes]

31. D - Information Security Management process is successfully implemented in an organization when all the following statements are true. I). Information is available as and when needed II). Information is secured with appropriate level of confidentiality III). Information is complete, accurate and is protected from unauthorized access IV). Information is exchanged between organization and others in a trusted, authenticated and reliable way. (Service Design) [ITIL Selected Processes]

32. C - The Goals of the Transition Planning and Support process in Service Transition phase of the Service Life Cycle are I). To ensure effective realization of Service requirements from Service Strategy and Service Design stages. II). Throughout the Service Transition activities, failure risks and disruptions are identified and managed. (Service Transition) [ITIL Selected Processes]

33. B - Short term improvements are those made to the working practices within the Service Operations processes, functions, and the technology itself. Generally they involve smaller improvements that do not change the fundamental nature of a process or technology, such as training, tuning, or personnel redeployments. [Service Operation]

34. B - The primary goal of Access management process is to protect the Security of the information assets and the IT infrastructure. (Service Operation) - Selected Processes [ITIL Selected Processes]

35. B - The Continual Service Improvement process feeds back to every stage in the Service Lifecycle. [ITIL Service Management as a practice]

36. C - The primary goal of the Service Operation is to deliver and support services. (Service Operation) - Selected Processes [ITIL Selected Processes]

37. A - The Change Manager issues the final decision with respect to the change. (Service transition) - Selected Processes [ITIL Selected Processes]

38. D - The objective of the Service Catalog Management process is to manage information within the Service Catalog and to ensure accuracy and quality prior to execution in the LIVE environment. (Service Design) [ITIL Selected Functions]

39. A - The purpose of Service Design stage in the Service Life Cycle is to design a new or changed service to be introduced into the LIVE environment. (Service Design) - Service Lifecycle [The Service Lifecycle]

40. B - If the services generate more value to the customer, this strengthens the business relations and the bond between the business and the customer. [ITIL Service management as a practice]

ITIL Foundation Practice Exam 13 Practice Questions

Test Name:
ITIL Foundation Practice Exam 13
Total Questions: 40
Correct Answers Needed to Pass: 30 (75.00%)
Time Allowed: 60 Minutes

Test Description

This is the eleventh cumulative ITIL Foundation test which can be used as an indicator for overall performance. This practice test includes questions from key ITIL areas.

Test Questions

1. Which of the following is the CORRECT order of the steps in the 7 Step Continual Improvement Process?
   I. Process the data
   II. Decide what can be measured
   III. Decide what should be measured
   IV. Gather the data

   A. III-II-IV-I

   B. IV-I-III-II

   C. II-III-IV-I

   D. I-IV-III-II

2. Which is the correct definition of a Configuration Item in ITIL?

   A. A Configuration Item (CI) is an asset, service component or other item which is, or will be, under the control of Configuration Management.

   B. A Configuration Item (CI) is a policy or rule which must be under the control of Configuration Management.

   C. A Configuration Item (CI) is an integral part of Configuration Management process which will help identify errors in the system.

   D. A Configuration Item (CI) is a document or a file which is, or will be, under the control of Configuration Management.

3. Which of the following statement/s is TRUE with respect to the Demand Management process?

   A. A demand can be technically managed (whether it is feasible or not) and financially managed (whether it is cost worthy or not).

B. A demand can only be financially managed(whether it is cost worthy or not).

C. A demand can only be technically managed(whether it is feasible or not).

D. A demand can be socially managed(whether it is allowed or not).

4. Identify the five elements of Information Security Management System Framework

I. Develop, implement, Control, Plan and Analyze

II. Develop, Control, Plan, Implement and Evaluate

III. Control, Plan, Implement, Evaluate and Maintain

IV. Control, Assess, Implement, Evaluate and Maintain

A. I

B. II

C. IV

D. III

5. Which of the following are NOT the activities as described in the Service Transition Plan for release and deploying a release into the Test and production Environments?

I. Service transition work environment and required infrastructure planning

II. Handling resource requirements and budgets at each Service Life Cycle stage

III. Identifying the test cases for the new service

IV. All other risks and issues to be controlled

A. IV

B. I

C. III

D. II

6. Which of the following statements is TRUE with respect to the definition of a Process?

A. A process takes the input and processes the output.

B. All processes need not have a process owner.

C. A process owner and a process manager always are two different roles and two different persons.

D. All processes should be measurable, cost and performance driven.

7. What are the contents of a Service Catalog?

A. A service Catalog contains a list of retired services provided by the Service Provider.

B. A service Catalog contains a list of decommissioned services provided by the Service Provider.

C. A service Catalog contains a list of live services provided by the Service Provider.

D. A service Catalog contains a list of live, retired and decommissioned services provided by the Service Provider.

8. Which of the following statements for a Service request is FALSE?

A. A service request is a request from a user for a standard change.

B. A service request is a request from a user for information or advice.

C. A service request is a request from a user for access to an IT service.

D. A service request is a request from a user for help him to open his software application.

9. Which of the following statements is NOT an example of an Incident?

A. A user calls the Service Desk to log a request to configure a new printer

B. A user submits a service desk call stating he is not able to open an application after an upgrade.

C. A user logs a call stating his inability to print.

D. A user logs a call to Service Desk stating his emails to the external domains are bouncing.

10. Which if the following is an example of a reactive change?

A. Adding more network bandwidth prior to a major business conference.

B. Replacing a failed network router that caused a service outage.

C. Upgrading a hard drive cluster that is 90% filled to capacity.

D. Virtualizing servers to save money.

11. Which of the following statements is a CORRECT definition of an Event in ITIL?

A. An event is a change of state that has significance for the management of a service in Service Catalog Management process.

B. An event is a change of state that has significance for the configuration management process.

C. An event is a change of process state that has significance for the IT management within an organization.

D. An event is a change of state that has significance for the management of a configuration item or IT service.

12. As per ITIL, which one of the following correctly defines the term Incident?

A. System failure with an unknown root cause.

B. Payroll System Security log showing many login failure incidents.

C. Any unplanned interruption to an IT SERVICE or reduction in the quality of an IT SERVICE.

D. Any event that does not form or constitute the standard operation of an IT Service or causes interruption to the regular service there by reducing the quality of service.

13. Which of the following is true with respect to the Service Portfolio?

I. The Service Catalog is a subset of the Service Portfolio

II. The Service Portfolio is a subset of the Service Catalog

A. None of these

B. II

C. I

D. Both I and II

14. Which of the following activities are NOT carried out by the Change Advisory Board(CAB)?

A. CAB assesses the changes.

B. CAB authorizes the changes.

C. CAB prioritizes the changes.

D. CAB executes the changes.

15. Which of the following statements are TRUE with respect to Management and Governance?

I. Management deals with decision making and process execution.
II. Governance deals primarily with decision making and not execution.
III. The terms Management and Governance are interchangeable.
IV. Management and Governance cannot survive together. Only one of them can be implemented in an organization.

A. I and II

B. IV

C. III

D. None of the statements are true

16. The entry to Incident Management process can come from which of the following?

I. Events communicated directly by users
II. Through Service Desk
III. Through tools used for event management
IV. Through hardware resources

A. II and IV

B. I and III

C. I, II and III

D. only II

17. Which of the following process has the ability to identify business priorities and allocate dynamic resources as and when needed?

A. Incident Management

B. Problem management

C. Change Management

D. Event Management

18. Which of the following definitions best apply to the Change models in the change Management process in the Service Transition phase?

A. Change models are steps to execute a change as defined in the Service Strategy phase of the Service Life cycle.

B. Change models reflect a fixed number and fixed type of steps required to do a change.

C. Change Models are defined as pre-established process flows with the necessary steps to satisfy the type of change and level of authorization needed to properly assess the risk and the impact.

D. Change models are templates used in a CMDB which are followed when updating Configuration Items (CIs).

19. The significance of RACI model is BEST described in which of the following statements?

A. Each activity in a process should have one person as Responsible owner and any number of persons as Accountable.

B. Each activity in a process should have only one person as Responsible owner and only one person as Accountable.

C. Each activity in a process should have at least one person as Responsible owner and only one person as Accountable.

D. Each activity in a process should have at least one person as Accountable and only one person as Responsible.

20. Which of the following activities does a Service Provider do at his best in order to achieve his business objectives?

A. Provide stable services to the customer.

B. Reacting to customer needs.

C. Predicting the customer needs through preparation and analysis of customer service usage patterns.

D. Enabling customer's business objectives.

21. Which of the following statements are FALSE with respect to the value provided by the Event Management Process to the business?

A. Event Management can help signal status changes that allow appropriate action to take place by manual intervention.

B. Event Management provides a basis for automated operations.

C. Event Management helps detect incidents at a very early stage.

D. Event Management cannot help in real-time monitoring to reduce downtime.

22. Which of the following is the aim of Availability Management?

I. To identify and resolve any kind of service related incidents and problems

II. To minimize the duration and impact of events on IT services to ensure rapid business execution

III. To support the effects of events on services for rapid business operation

IV. To examining the events effected on services for rapid business execution

A. I only

B. II only

C. III only

D. IV only

23. Please state which of the following choices of answers best suit the given statement.
Service Improvements will be possible only if there exists

I. Well planned and implemented processes

II. Performance is monitored on a day-to-day basis

III. Metrics are gathered appropriately

IV. Data is gathered systematically

A. All of them

B. II, III and IV

C. I, II and III

D. I, III and IV

24. Which are the two major control perspectives of process guidelines, methods and tools with respect to IT Service Operation phase of the Service Life Cycle?

A. Demand and Supply

B. Manage and Support

C. Reactive and Proactive

D. Good practice and Best practice

25. Which of the following choices correctly depict the sub-types in the Problem Management process?

A. Proactive and Reactive

B. Subjective and Objective

C. Customer oriented and Business oriented

D. Internal and External

26. What are the service provider types according to the Service Strategy phase?

A. Internal, Shared and External

B. Internal and External

C. Internal and Outsourced

D. Shared and External

27. What is the definition of a "Function" in ITIL?

I. Units of organization specialized to perform certain types of work
II. Units of organization responsible for specific outcomes
III. Units of organization self contained with capabilities and resources required to perform and produce the result with BOK

A. I and II only

B. I, II and III

C. I only

D. I and III only

28. Event Management can be applied to which of the following aspects of Service management?

I. Configuration items
II. Environmental conditions
III. Software license monitoring
IV. Security

A. II, III and IV

B. I, II and III

C. All of them

D. I, II and IV

29. The Service V Model helps us to achieve which of the following objectives?

A. It helps us to understand the customer requirements better.

B. It helps us understand the validity of quality measurements for a given service.

C. It helps us to understand the difference between the projected business cost and the actual business costs.

D. It helps us to understand the verification and validation test requirements of the service.

30. Which of the following choices correctly define the term KEDB?

A. Knowledge Executive Draft Board

B. Known Error Database

C. Known Event Database

D. Knowledge Error Database

31. Which one of the following is NOT one of the delivery models in the Service Design Life cycle stage?

A. ASP

B. KPO

C. BPO

D. NET

32. Which of the following statements are TRUE with respect to the Information Security Management process?

I. Business will decide which information is to be classified as confidential and which is to be public

II. All business processes except Service improvement processes should be considered when formulating the level of data protection.

III. Information and Data protection involves physical and technical aspects.

A. II and III

B. All of the statements are true

C. I and II

D. I and III

33. What are the contents of a Definitive Media Library as suggested by ITIL?

A. Master copies of all controlled documentation

B. Licenses of all applications and software

C. Master copies of all developed software application

D. All of them

34. Which of the following choices BEST describe a Service Desk structure in ITIL?

A. Local, Centralized, Virtual, Follow the sun

B. Local, Centralized, Real, Outsourced

C. Local, International, Outsourced, Under the sun

D. Physical, Virtual, National, International

35. Which of the following are the goals of the Change Management Process in the Service Transition phase?

I. The Configuration Management System handles all changes to Configuration items and Service assets

II. The goal of change management process is in alignment with the business goals and all stakeholders' interest

III. Prevent unauthorized access to people to make changes to the production environment

A. I and III

B. I and II

C. II and III

D. I, II and III

36. Which of the following are the types of change models as per the Change Management process?

A. Standard, Normal and Emergency

B. Standard, Classified and Emergency

C. Normal, Average and Emergency

D. Normal, Emergency and Urgent

37. Which of the following are NOT activities related to the Service Strategy process?

I. Survey the market and understand the needs of the market.

II. Articulate the needs of the customer and define the value for your service in his terms.

III. Analyze the business patterns and prescribe the way in which they will access the service over a period of time.

IV. Agreeing upon service level targets with the customer after negotiation to produce the SLA.

A. III and IV

B. Only II

C. I and II

D. I and III

38. Which of the following are FALSE with respect to the Change Management Process in the Service Transition phase?

I. Increase the mean time to restore service (MTRS) by quickly and successfully implementing the corrective changes

II. Track changes through the Service Lifecycle on an on-and-off basis

III. Work towards poorer estimations of the quality, time and cost of change

IV. Assess the business risks and risks to the customer that are associated with the transition of services

A. only III and IV

B. I, II and III

C. All of the statements are false

D. I, II and IV

39. What are the two main characteristics of Service assets?

A. Outsourced and rented

B. Internal and External

C. Shared and Exclusive

D. Fit for use and Fit for purpose

40. What are the two main characteristics of the Service assets?

A. Shared and Exclusive

B. Utility and Warranty

C. Outsourced and rented

D. Internal and External

ITIL Foundation
Practice Exam 13
Answer Key and Explanations

1. A - The following is the CORRECT order of the steps in the 7 Step Continual Improvement Process. Decide what should be measured , Decide what can be measured, Gather the data and Process the data. (Continual Service Improvement) - Selected Processes [ITIL Selected Processes]

2. A - The correct definition of a Configuration Item in ITIL) is - "A Configuration Item (CI) is an asset, service component or other item which is, or will be, under the control of Configuration Management." (Service Transition) - Selected Processes [ITIL Selected Processes]

3. A - A demand can be technically managed(whether it is feasible or not) and financially managed(whether it is cost worthy or not). (Service strategy) - Selected Processes [ITIL Selected Processes]

4. D - Control, Plan, Implement, Evaluate and Maintain are the five elements of Information Security Management System Framework. (Service Design) - Selected Processes [ITIL Selected Processes]

5. C - Identifying the test cases for the new service is not an activity as described under in the Service Transition Plan for release and deploying a release into the Test and production Environments. (Service Transition) - Selected Processes [ITIL Selected Processes]

6. D - All processes should be measurable, cost and performance driven. (ITIL Generic Concepts) - Generic Concepts and Definitions [ITIL Generic Concepts and Definitions]

7. C - A service Catalog contains a list of live services provided by the Service Provider. (Service Design) - Selected Processes [ITIL Selected Processes]

8. D - A request from a user for help him to open his software application is an Incident, others are service requests. (Service Operation) - Selected Processes [ITIL Selected Processes]

9. A - A user calls Service Desk to log a request to configure a new printer - This is a Service request and not an Incident. (Service Operation) - Selected Processes [ITIL Selected Processes]

10. B - Reactive changes are implemented in response to an incident, problem, or event, while proactive changes are implemented in advance of a predicted need. [Service Transition] [ITIL Technology and Architecture]

11. D - The CORRECT definition of an Event in ITIL) is "An event is a change of state that has significance for the management of a configuration item or IT service." (Service Operation) - Selected Processes [ITIL Selected Processes]

12. C - ITIL) defines an Incident as "An unplanned interruption to an IT SERVICE or reduction in the quality of an IT SERVICE". (Service Operation) - Generic Concepts and Definitions [ITIL Generic Concepts and Definitions]

13. C - The Service Catalog is a subset of the Service Portfolio. This is a key concept of the Service Design life cycle phase. (Service Design) - Service Lifecycle [The Service Lifecycle]

14. D - CAB Does not execute the change. It only supports the Change Management team appropriately by assessing, prioritizing and authorizing the changes. (Service Transition) - Selected Processes [ITIL Selected Processes]

15. A - The following statements are TRUE with respect to Management and Governance.
I). Management deals with decision making and process execution.
II). Governance deals with only decision making and no execution. (ITIL Generic Concepts) - Generic Concepts and Definitions [ITIL Generic Concepts and Definitions]

16. C - The entry to Incident Management process can come from Events communicated directly by users, Through Service Desk and Through tools used for event management. (Service Operation) - Selected Processes [ITIL Selected Processes]

17. A - Incident management process has the ability to identify business priorities and allocate dynamic resources as and when needed. (Service Operation) - Selected Processes [ITIL Selected Processes]

18. C - The definition of the Change models as present in the change Management process in Service Transition phase is "Change Models are pre-established process flows with the necessary steps to satisfy the type of change and level of authorization needed to properly assess risk and impact". (Service Transition) -

Selected Processes [ITIL Selected Processes]

19. C - Each activity in a process should have at least one person as Responsible owner and only one person as Accountable. (ITIL Generic Concepts) - Generic Concepts and Definitions [ITIL Generic Concepts and Definitions]

20. C - Predicting the customer needs through preparation and analysis of customer service usage patterns is the only way a service provider can excel in his job. (ITIL Generic Concepts) - Generic Concepts and Definitions [ITIL Generic Concepts and Definitions]

21. D - Event Management can help significantly in real-time monitoring to reduce downtime by automating activity. (Service Operation) - Selected Processes [ITIL Selected Processes]

22. B - The Aim of the Availability Management is to minimize the duration and impact of events on IT services to ensure rapid business execution. (Service Design) - Selected Processes [ITIL Selected Processes]

23. A - Service Improvements will be possible only if there exists
I). Well planned and implemented processes
II). Performance is monitored on a day-to-day basis
III). Metrics are gathered appropriately
IV). Data is gathered systematically.
(Continual Service Improvement) - Selected Processes [ITIL Selected Processes]

24. C - The two major control perspectives of process guidelines, methods and tools with respect to IT Service Operation phase of the Service Life Cycle are Reactive and Proactive. (Service Operation) - Generic Concepts and Definitions [ITIL Generic Concepts and Definitions]

25. A - Proactive and Reactive correctly depict the sub-types in the Problem Management process. (Service Operation) - Selected Processes [ITIL Selected Processes]

26. A - Internal, Shared and External are the service provider types according to the Service Strategy phase. (Service Strategy) - Service Lifecycle [The Service Lifecycle]

27. B - A Function in ITIL is defined as
I). Units of organizations specialized to perform certain types of work
II). Units of organizations responsible for specific outcomes

III). Units of organization self contained with capabilities and resources required to perform and produce the result with BOK

(ITIL Generic concepts) - Generic Concepts and Definitions [ITIL Generic Concepts and Definitions]

28. C - Event Management can be applied to all the following aspects of Service management.

I). Configuration items

II). Environmental conditions.

III). Software license monitoring.

IV). Security.

(Service Operation) - Selected Processes [ITIL Selected Processes]

29. D - The Service V Model helps us to understand the verification and validation test requirements of the service. (Service Transition) - Key Principles and Models [ITIL Key Principles and Models]

30. B - KEDB correctly expands as Known Error Database.(Service Operation) - Selected Processes [ITIL Selected Processes]

31. D - BPO, KPO & ASP are the delivery models in the Service Design Life cycle stage. .Net is an unrelated platform reflecting a programming technology. (Service Design) - Service Lifecycle [The Service Lifecycle]

32. D - The following statements are TRUE with respect to the Information Security management process

I). Business will decide which information is to be classified as confidential and which is to be public

II). Information and Data protection involves physical and technical aspects. (Service Design) - Selected Processes [ITIL Selected Processes]

33. D - The contents of a Definitive Media Library as suggested by ITIL are master copies of all applications, Licenses and controlled documentation. (Service Transition) - Selected Processes [ITIL Selected Processes]

34. A - Local, Centralized, Virtual, Follow the sun are the Service Desk structured in ITIL). (Service Operation) - Selected Functions [ITIL Selected Functions]

35. D - The following are the goals of the Change Management Process in the Service Transition phase

I). The Configuration Management System handles all changes to Configuration items and Service assets.

II).The goal of change management process is in alignment with the business goals and all stakeholders' interest.

III). Prevent unauthorized access to people to make changes to the production environment. (Service Transition) - Selected Processes [ITIL Selected Processes]

36. A - The following are the types of change models as per the Change Management process - Standard, Normal and Emergency. (Service Transition) - Selected Processes [ITIL Selected Processes]

37. A - Service Strategy Process is concerned with strategizing the conception of the service as per the market demand, development of the service, planning for the resources and execution. The following are some of the key activities: Survey the market and understand the needs of the market; Choose your customers and zero in on them; Articulate the needs of the customer and define the value for your service in his terms; Define the kinds of services as a service provider you can provide to the customer; To top up the service define the value add you can provide to your customers (Service Strategy) - Selected Processes [ITIL Selected Processes]

38. B - All choices except "Assessing the business risks and risks to the customer that are associated with the transition of services" are FALSE with respect to the Change Management Process in the Service Transition phase. (Service Transition) - Selected Processes [ITIL Selected Processes]

39. D - The two main characteristics of Service assets are Fit for purpose and Fit for use. (Service Strategy) - Generic Concepts and Definitions [ITIL Generic Concepts and Definitions]

40. B - The two main characteristics of Service assets are Utility and Warranty. (Service Strategy) - Service Lifecycle [The Service Lifecycle]

Knowledge Area Quiz: More Selected Processes Practice Questions

Test Name:
Knowledge Area Quiz: Selected Processes
Total Questions: 20
Correct Answers Needed to Pass:
14 (70.00%)
Time Allowed: 20 Minutes

Test Description

This extended practice test targets ITIL concepts related to Selected Processes.

Test Questions

1. Which one of the following is the primary goal of the Service Portfolio Management process?

   A. To support the service management process by managing the information storage and access.

   B. To manage the suppliers and their services which in turn help the provider in running his services with an ultimate goal of value for money.

   C. To articulate the business needs and the provider's response to those needs.

   D. To articulate the IT support needs and the Portfolio Management process.

2. Which of the following statements are TRUE with respect to the Service Catalog?
   I. Service Catalog speaks of the actual and present capabilities of the service provider.
   II. Service Catalog enables the service provider to customize service solutions for any customer.

   A. None of these statements are true

   B. I and II

   C. II only

   D. I only

3. In which stage does the Service Catalog Management act as the key process and important activity of the Service Life Cycle?

   A. Service Strategy

   B. Service Transition

C. Service Design

D. Service Operation

4. Which of the following are the Key Processes as defined by Continual Service Improvement process for effective implementation of Continual Improvement?

I. Service Interface
II. 7-Step Improvement Process
III. Service Measurement
IV. Service Reporting

A. II, III and IV

B. All of these items are key processes of CSI

C. I, III and IV

D. I, II and III

5. Which two of the following are variants of the Service Catalog?

I. Business Service Catalog
II. IT Service Catalog
III. Technical Service Catalog
IV. Marketing Service Catalog

A. I and II

B. II and IV

C. III and IV

D. I and III

6. Which of the following are the key activities within the Service Catalog Management process?

I. Interfacing with internal support teams and Suppliers and dependencies between IT services and the supporting services.
II. Agreeing upon a common definition for service with all relevant parties concerned.
III. Interfacing with Service Portfolio Management to agree upon the contents of the Service Portfolio and Service Catalog.

A. II and III only

B. I and III only

C. I, II and III

D. I and II

7. Which of the following statements are TRUE with respect to the SLA?

I. The SLA is effectively a service quality level warranty or assurance by the service provider.

II. The SLA is considered for each of the services provided by the service provider.
III. The success of the SLA determines the quality of the Service Portfolio and the Service Catalog.

A. None of these statements are true

B. I and III only

C. II and III only

D. I, II and III

8. Which of the following is NOT an activity of the Service Level Management Process?
I. Design of appropriate services, technology, processes, information and process measurements to meet the business requirements
II. Produce Service Reports and maintain documents related to SLM standards.
III. Develop and maintain relationships with all concerned stakeholders of service provision.
IV. Record and manage all complaints and compliments.

A. I only

B. I and III

C. II and IV

D. II only

9. Which of the following are used to judge the effectiveness and efficiency of SLM activities?
I. KPI
II. Metrics
III. SIP

A. I and II only

B. I, II and III

C. None of these items are used to judge the effectiveness and efficiency of SLM activities

D. I and III only

10. Which of the following is NOT an objective of the KPI in the service design phase?
I. Number or percentage of service targets being met
II. Number of incidents per month
III. Number of services with up-to-date SLAs
IV. Number of services with timely reports and active service reviews

A. I, II

B. I, II, IV

C. II only

D. III, IV

11. Which stage of the Service Life Cycle evaluates the capabilities of current stage of the Service delivery?

A. Service Transition

B. Service Strategy

C. Service Design

D. Service Catalog

12. Which stage of the Service Life Cycle evaluates the existing Service Catalog to determine whether existing and live services can be coupled to deliver the required business results?

A. Service Transition

B. Service Design

C. Service Catalog

D. Service strategy

13. Which of the following are the key Service Transition processes?

I. Service activation Planning and Support
II. Change Management
III. Service Validation and Testing Management
IV. Release and Deployment Management

A. II, III and IV

B. only III and IV

C. I, III and IV

D. only I and II

14. The Service Life Cycle stage - Continuous Service Improvement is all about looking for ways to do which of the following?

A. Improve employee training

B. Improve cost effectiveness

C. Improve process effectiveness

D. Improve process efficiency

15. Which of the following are the objectives of Continuous Service Improvement Process?

I. Review and analyze each phase of the service life cycle and identify and recommend improvement opportunities
II. Review and analyze the results of Service Level achievement
III. Keeping customer satisfaction in mind, deliver IT services cost effectively

A. All of these items are objectives of CSI

B. II and III only

C. I and III only

D. I and II only

16. Periodically conducting internal audits verifying employee and process compliance'. This activity belongs to which stage of the Service Life Cycle?

A. Service Design

B. Service Operation

C. Service Transition

D. Continuous Service Improvement

17. Which of the following statements is TRUE with respect to SFA in Service Design?

I. SFA is a structured technique to identify improvement opportunities in end-to-end service availability that can deliver benefits to the user.
II. SFA helps identify and analyze the root cause of service interruptions

A. None of these items are true

B. II only

C. I only

D. Both I and II

18. Availability Management process deals with methods and techniques that help one or more of these

I. Analyze service failure
II. Prevent service failure
III. Assess service failure

A. I, II and III

B. II only

C. III only

D. I only

19. Which one of the following statements is the Primary goal of the

IT Service Continuity Management process

I. To design and develop the processes and services that continue to operate
II. To design the services so that they can resume operations as soon as possible
III. To ensure that the services never fail
IV. To restore normalcy of the IT Service Delivery within agreed business timescales

A. III

B. II

C. I

D. IV

20. Service continuity is implemented and managed in which of these four stages?

I. Initiation
II. Requirements and strategy
III. Assessment and analysis
IV. Implementation
V. Ongoing operation

A. I, II, III and V

B. I, II, III and IV

C. I, II, IV and V

D. II, III, IV and V

Knowledge Area Quiz:
More Selected Processes
Answer Key and Explanations

1. C - The primary goal of the Service Portfolio Management is to articulate business needs and the provider's response to those needs. (Service Strategy) - Selected Processes [ITIL Selected Processes]

2. B - Both statements are TRUE regarding the Service Catalog; as it reflects the actual and present capabilities of the service provider, and enables the service provider to customize service solutions for any customer. (Service Design) - Selected Processes [ITIL Selected Processes]

3. C - Service Catalog management Process belongs to the Service Design stage of the Service Life Cycle. (Service Design) - Selected Processes [ITIL Selected Processes]

4. A - 7-Step Improvement Process, Service Measurement and Service Reporting are the key processes as defined by Continual Service Improvement process for effective implementation of Continual Improvement. (Continual Service Improvement) - Selected Processes [ITIL Selected Processes]

5. D - Business Service Catalog and Technical Service Catalog are the two variants of the Service Catalog. (Service Design) - Selected Processes [ITIL Selected Processes]

6. C - Key activities within the Service Catalog Management Process include interfacing with internal support teams and suppliers; agreeing upon a common definition for service with all relevant parties concerned; and interfacing with Service Portfolio Management process to agree upon the contents of the Service Portfolio and Service Catalog. - Selected Processes [ITIL Selected Processes]

7. D - The following statements are true with respect to the SLA.
   I). The SLA is effectively a service quality level warranty or assurance by the service provider.
   II). The SLA is considered for each of the services provided by the service provider.
   III). The success of the SLA determines the quality of the Service Portfolio and the Service Catalog. (Service Design) - Selected Processes [ITIL Selected Processes]

8. A - The following are the correct activities of SLM process.

I). Produce Service Reports and maintain documents related to SLM standards.
II). Develop and maintain relationships with all concerned stakeholders of service provision.
III). Record and manage all complaints and compliments. (Service Design)
- Selected Processes [ITIL Selected Processes]

9. B - Key Performance Indicators (KPIs), Metrics and Service Improvement Plans (SIPs) are used to judge the effectiveness and efficiency of SLM activities. (Service Design) - Selected Processes [ITIL Selected Processes]

10. C - Number of incidents per month is NOT an objective of KPI in Service Design phase. The objectives are
I). Number or percentage of service targets being met
III). Number of services with up-to-date SLAs
IV). Number of services with timely reports and active service reviews. (Service Design) - Selected Processes [ITIL Selected Processes]

11. B - The Service Strategy stage of the Service Life Cycle evaluates the capabilities of current stage of the Service delivery. (Service Strategy) - Selected Processes [ITIL Selected Processes]

12. B - The Service Design stage of Service Life Cycle evaluates the existing Service Catalog to determine whether existing and live services can be coupled to deliver the required business results. (Service Design) - Selected Processes [ITIL Selected Processes]

13. A - The key Service Transition processes are Change Management, Service Validation and Testing Management and Release and Deployment Management. (Service Transition) - Selected Processes [ITIL Selected Processes]

14. C - The Service Life Cycle stage - Continuous Service Improvement is all about looking for ways Improve process effectiveness, Improve process efficiency and Cost effectiveness. (Continuous Service Improvement) - Selected Processes [ITIL Selected Processes]

15. D - These are the objectives of Continuous Service Improvement Process.
I). Review and analyze each phase of the service life cycle and identify and recommend improvement opportunities

II). Review and analyze the results of Service Level achievement. (Continual Service Improvement)
- Selected Processes [ITIL Selected Processes]

16. D - Periodically conducting internal audits verifying employee and process compliance' activity belongs to the Continuous Service Improvement stage of the Service life Cycle. (Continual Service Improvement) - Selected Processes [ITIL Selected Processes]

17. D - SFA is a structured technique to identify improvement opportunities in end-to-end service availability that can deliver benefits to the user.
SFA also helps identify and analyze the root cause of service interruptions. (Service Design) - Selected Processes [ITIL Selected Processes]

18. A - Availability Management process deals with methods and techniques that help Analyze service failure, Prevent service failure and Assess service failure. (Service Design) - Selected Processes [ITIL Selected Processes]

19. D - The Primary goal of IT Service Continuity Management is to restore normalcy of the IT Service Delivery within agreed business timescales. (Service Design) - Selected Processes [ITIL Selected Processes]

20. C - Service continuity is implemented and managed in the following four stages:
Initiation, Requirements and strategy, Implementation and Ongoing operation. (Service Design)
- Selected Processes [ITIL Selected Processes]

ITIL Foundation Practice Exam 14 Practice Questions

Test Name:
ITIL Foundation Practice Exam 14
Total Questions: 40
Correct Answers Needed to Pass:
30 (75.00%)
Time Allowed: 60 Minutes

Test Description

This is a cumulative ITIL Foundation test which can be used as a baseline for initial performance. This practice test includes questions from all ITIL question categories.

Test Questions

1. The Availability Manager is responsible for achieving what level of availability?

   A. Availability within +/- 5% of SLA terms.

   B. 100% uptime

   C. Availability equal to or greater than SLA terms

   D. 99.999% uptime during business hours

2. One of Service Operation's key goals is achieving balance between the Internal IT View and the External Business View, as well as which of the following?

   A. Stability and responsiveness

   B. Budget and requirements

   C. Availability and capacity

   D. Incidents and problems

3. The finance group runs a large database report daily. This report is very resource intensive and slows the response time for all users of the database while the report is running. After some discussion with the IT group, it was determined that running the report after business hours would be the most effective way to resolve the problem of slow database performance for all users. What is this response called?

   A. Demand management

   B. Availability management

   C. Capacity management

   D. Service management

4. Which Service Design process is concerned with negotiating and establishing performance targets?

   A. Service Level Management

B. Availability Management

C. Capacity Management

D. IT Service Continuity Management

5. Which of the following processes is focused on the underlying IT infrastructure?

A. Service Capacity Management

B. Component Capacity Management

C. Demand Management

D. Availability Management

6. Leggett Accounting has found that its corporate mail system has consistently fallen below stated levels of availability. Which of the following would be developed to determine how best to improve email services?

A. Service Improvement Plan

B. Availability Management Plan

C. Service Metric Analysis

D. Capacity Management Plan

7. Financial Management for IT is comprised of which three activities?

A. Budgeting, IT accounting, and demand management

B. Budgeting, IT accounting, and charging

C. Budgeting, opportunity management, and charging.

D. Budgeting, analysis, and charging

8. Which of the following are inputs into Capacity Management processes?

A. Budgets

B. Business strategies

C. Business requirements

D. All of the above

9. ITIL separates which of the following into these groups: Process, Technology, and Service?

A. Metrics

B. Business requirements

C. Contracts

D. Personnel

10. Which ITIL lifecycle process is most concerned with disaster recovery planning?

A. IT Service Continuity Management

B. Availability Management

C. Capacity Management

D. Risk Management

11. Key Performance Indicators are metrics that provide an objective view of a process's performance. The number of successful changes implemented in a 6 month period is a measure of which process?

A. Capacity Management

B. Change Management

C. Configuration Management

D. Availability Management

12. Ted is reviewing the processes and activities of the sales department over time, in order to predict future demand for the service that he manages, and discovers activity levels typically increase at month-, quarter-, and year-end. What this set of metrics he is reviewing called?

A. Patterns of business activities

B. Service level agreements

C. Availability reports

D. Capacity reports

13. Predicting and balancing capabilities, resources, and demand are the goals of which of the following processes?

A. Demand Management

B. Capacity Management

C. Component Management

D. Service Management

14. Johnson Power Controls is conducting a Configuration Audit. Which of the following could be a catalyst for this audit?

A. Corporate policy requiring audits at regular intervals

B. All of the above

C. Discovery of an unauthorized CI

D. Significant changes to the IT infrastructure

15. Acme Systems has contracted with Payroll Pro to handle all their payroll duties. What is type of arrangement called?

A. Partnership

B. Co-sourcing

C. Shared Services

D. Business Process Outsourcing

16. Which type of SLA offers the most flexibility and efficiency for an organization?

A. Underpinning Contract

B. Service-based

C. Multi-level

D. Customer-based

17. With which of the following groups must the Change Management Team most closely coordinate?

A. Customers

B. IT Staff

C. Executives

D. None of the above

18. Johnson Enterprises' IT team has determined that an additional storage array will be required to support a new company database. What is this activity called?

A. Performance Monitoring

B. Application Sizing

C. Modeling

D. Demand Management

19. The Service Catalog contains an organization's Business Services Catalog and Technical Service Catalog, both of which contain details of all the IT services delivered to the customer, each from a different perspective. The Business Service Catalog provides what view of the Service Catalog?

A. Shared Services view

B. Contract view

C. Customer view

D. Departmental view

20. Which of the following best describe the phases of an incident between occurrence and restore point?

A. Detection, Diagnosis, Repair, Recovery

B. Monitoring, Measuring, Reporting, Recovery

C. Detection Diagnosis, Notification, Recovery

D. Detection, Management, Reporting, Analysis

21. Deepak is the Service Level Manager for a cloud services provider. Which of the following is not one of his responsibilities:

A. Developing Underpinning Contracts

B. Vendor relationship management

C. Developing Service Level Agreements

D. Service Testing

22. The Change Manager at Phillips Morgan Accounting does not have the technical expertise to understand the possible impacts of a proposed change to a database server. According to ITIL, upon which of the following would the Change Manager rely to provide advice and guidance?

A. Change Advisory Board

B. Finance Director

C. IT Director

D. Suppliers

23. Donna is the Change Manager at ABC Manufacturing. In order for her to provide final authorization for a change, she must have approval from which three organization areas?

A. Suppliers, Business, Technology

B. Finance, Executive, Technology

C. CAB, ECAB, Configuration Management

D. Finance, Business, Technology

24. A service provider wants to identify the roles of the stakeholders of a new product it will offer. Which of the following models would be useful for this purpose?

A. Flowchart

B. Swim-lane

C. FMIT

D. RACI

25. Which of the following types of support models offers the highest first call resolution rate?

A. Call Center

B. Service Desk

C. Help Desk

D. Customer Service Desk

26. Echo Systems has setup databases and tools to capture facts and statistics about its IT services. Which of the following can not be captured this way?

A. Wisdom

B. Information

C. Knowledge

D. Data

27. Ensuring the confidentiality, integrity, and availability of an organizations IT assets is the goal of which ITIL lifecycle process?

A. Information Security Management

B. IT Service Management

C. IT Service Continuity Management

D. Availability Management

28. ITIL describes a model for ensuring a service is ready for release to customers that involves specifying and validating all requirements. What is this model called?

A. Matrix

B. Iterative

C. Waterfall

D. Service V

29. Richard maintains a large inventory of spare parts for the systems in his company's IT infrastructure. These spares include disk drives, network switches, and VoIP phones, all of which are pre-configured and ready to be put into production. What is this inventory called?

A. Hardware Library

B. Definitive Spares

C. Just In Time Inventory

D. Hot Spares

30. Plan, Do, Check, Act are the four phases of which continuous process improvement model?

A. Ishikawa Model

B. ITIL

C. Deming Cycle

D. Maslow's Hierarchy

31. Which of the following is not one of the four perspectives of ITSM?

A. People

B. Partners

C. Policy

D. Products

32. Which of the following is not a type of control used by Information Security Management?

A. Technical

B. Operational

C. Physical

D. Procedural

33. Separating file backup and file restore tasks between staff members is an example of what type of security control?

A. Organizational

B. Physical

C. Procedural

D. Technical

34. Solis Enterprises experienced a 45-minute outage on its primary internet connection during core business hours. The network group manually transitioned the company's internet access to a slower speed backup connection until the primary circuit came back online. Which process best describes this response?

A. Event Management

B. Availability Management

C. Incident Management

D. Problem Management

35. Any time a new service is developed or an existing service is modified, it must be checked against which of the following?

A. Processes

B. Technology architectures

C. Business requirements

D. Service pipeline

36. Mary Beth is the Release and Deployment Manager at Stepco Design. She interfaces with the Change Management Team as well as various engineering and infrastructure groups developing new services. Which of the following skills would Mary Beth most need to be successful in this role?

A. Project Management

B. Public Relations

C. Business Awareness

D. Budgeting

37. A new or modified service that is released to all users is known as what type of deployment?

A. Push

B. Attended

C. Automated

D. Big Bang

38. Monitoring, measuring, and analyzing the performance of a server with the intent to alert IT an outage has occurred is which type of activity?

A. Proactive

B. Reliability management

C. Outage management

D. Reactive

39. An organization has determined a service no longer meets minimum functional and technical requirements. Which of the following outcomes should the organization choose for this service?

A. Replace

B. Retire

C. Renew

D. Refactor

40. Which of the following processes is found in two ITIL Lifecycle phases?

A. Problem Management

B. Requirements Management

C. Service Level Management

D. Capacity Management

ITIL Foundation Practice Exam 14 Answer Key and Explanations

1. C - The Availability Manager is responsible for achieving levels of availability that meet or exceed business requirements - ITIL - Selected Roles [ITIL Selected Roles]

2. A - An appropriate balance must be struck between stability and responsiveness. Too much focus on stability can cause IT to ignore changing business requirements, while too much focus on responsiveness can result in a decrease in the stability of the infrastructure. - ITIL - Service Management as a Practice [Service Management as a Practice]

3. A - Demand management is the process of understanding business demands for services, and strategically responding with physical, technical, or financial constraints to influence and control the demand. - ITIL - Selected Processes [ITIL Selected Processes]

4. A - Service Level Management concerns itself with determining, negotiating, and establishing service delivery target levels. - ITIL - Selected Processes [ITIL Selected Processes]

5. B - Component Capacity Management provides for the identification and management of IT infrastructure components such as CPU, memory, bandwidth, and disk utilization. - ITIL - Selected Processes [ITIL Selected Processes]

6. A - Service Improvement Plans are used to implement necessary changes to improve a service. A baseline measurement is used as a reference point to determine if improvement goals have been met. - ITIL - Selected Processes [ITIL Selected Processes]

7. B - Budgeting, IT accounting, and charging are the primary activities performed in Financial Management for IT. - ITIL - Selected Processes [ITIL Selected Processes]

8. D - Capacity management uses inputs from both business and technical camps to ensure adequate and appropriate performance and capacity of IT services and components. - ITIL - Selected Processes [ITIL Selected Processes]

9. A - ITIL categorizes metrics as either Process, Technology, or Service. Process Metrics are captured as KPIs, Technology Metrics are associated with infrastructure and applications, and Service Metrics measure the end to end performance of a service by combining multiple component metrics. - ITIL - Service Management as a Practice [Service Management as a Practice]

10. A - IT Service Continuity Management ensures that the required IT infrastructure and services that support critical business processes can be recovered within required

timeframes. - ITIL - Selected Processes [ITIL Selected Processes]

11. B - The number of changes implemented is a KPI for Change Management. This metric could be further analyzed to determine the number or percentage of successful changes. - ITIL - Selected Processes [ITIL Selected Processes]

12. A - Business processes are the primary source of demand for services. Analysis of the patterns of business activity allows a service provider to predict, strategically plan for, and respond to changes in demand for supporting services. - ITIL - Selected Functions [ITIL Selected Functions]

13. B - Capacity Management provides predictive capacity indicators and balances resources, capabilities, and demand. - ITIL - Selected Processes [ITIL Selected Processes]

14. B - Configuration Audits verify that CI's exist and they are correctly recorded. It is appropriate to conduct an audit anytime there is a change in the item's operating environment, when dictated by corporate policy, or when unauthorized CI's are discovered. - ITIL - Selected Processes [ITIL Selected Processes]

15. D - Using a 3rd party to provide and manage one or more of an organization's business processes, such as payroll or logistics, is called Business Process Outsourcing. The management of this arrangement is a function of the Supplier Management process. - ITIL - Selected Functions [ITIL Selected Functions]

16. C - Multi-level SLAs permit an organization to customize services and service offerings, while minimizing the effort required to do so. - ITIL - Key Principles and Models [ITIL Key Principles and Models]

17. D - To be most effective, the Change Management process must be impartial to all groups within an organization. This impartiality allows the Change Management team to make decisions that best support the organization as a whole. - ITIL - Selected Roles [ITIL Selected Roles]

18. B - Application sizing focuses on determining the hardware or network capacity required to support an application and its predicted usage. - ITIL - Selected Functions [ITIL Selected Functions]

19. C - The Business Service Catalog contains the relationships between business units and business processes that rely on IT services, and is the customer view of the Service Catalog. - ITIL - Selected Functions [ITIL Selected Functions]

20. A - The phases of the incident lifecycle are Detection, Diagnosis, Repair, and Recovery. Time metrics are used to measure these phases. - ITIL - Key Principles and Models [ITIL Key Principles and Models]

21. D - Service Level Managers are responsible for developing and managing service level agreements as well as the supporting vendor contracts the service requires. - ITIL - Selected Roles [ITIL Selected Roles]

22. A - The Change Advisory Board (CAB) is the body of experts that the Change Manager relies on for advice and information when changes are proposed. The CAB is composed of representatives from various stakeholder groups such as IT, Finance, HR, or even 3rd party suppliers. - ITIL - Selected Roles [ITIL Selected Roles]

23. D - Although the Change Manager is responsible for authorizing Changes, he or she must have Financial, Business, and Technology approval before final approval is granted. - ITIL - Selected Roles [ITIL Selected Roles]

24. D - A RACI model is used to document roles and relationships between stakeholders. Stakeholders are identified as one or more of the following: Responsible, Accountable, Consult, and Inform. - ITIL - Key Principles and Models [ITIL Key Principles and Models]

25. B - The Service Desk offers the highest rate of first-call resolution for users' requests for support. Service desks are generally staffed by agents with a wide range of communication and technical skills to best respond to various types of service requests. - ITIL - Selected Functions [ITIL Selected Functions]

26. A - Wisdom comes from having knowledge and experience to make sound judgments and decisions. It is a quality that can not be captured in a database or tool. - ITIL - Key Principles and Models [ITIL Key Principles and Models]

27. A - Information Security Management ensures that the confidentiality, integrity, and availability of an organization's assets, including data, services, and infrastructure, is maintained in accordance with the organization's policies. - ITIL - Selected Functions [ITIL Selected Functions]

28. D - The Service V Model is a structured approach to defining acceptance requirements against requirements for features, functionality, performance. When the acceptance requirements for all aspects of the service have been met, the service is deemed ready for release to customers. - ITIL - Generic Concepts and Definitions [ITIL Generic Concepts and Definitions]

29. B - Spare equipment and components that are preconfigured and maintained at the same level as the production environment are housed in an inventory called Definitive Spares (DS). - ITIL - Technology and Architecture [ITIL Technology and Architecture]

30. C - The Deming Cycle, also known as PDCA, is a continuous process improvement model. Each step is carried out in this specific order, as many times as necessary. - ITIL - Key Principles and Models [ITIL Key Principles and Models]

31. C - The four perspectives of IT service management, also known as "4P's", are partners, people, products, and processes. - ITIL - Service Management as a Practice [Service Management as a Practice]

32. B - Information Security Management uses organizational, procedural, physical, and technical controls to secure IT assets. - ITIL - Selected Functions [ITIL Selected Functions]

33. C - Separation of duties between two or more staff members is a procedural security control. - ITIL - Selected Functions [ITIL Selected Functions]

34. C - Incident Management addresses the symptoms of an unplanned disruption or outage to restore service operation as quickly as possible. It does not address the root cause of the issue. - ITIL - Selected Processes [ITIL Selected Processes]

35. C - An organization must cross check every new and modified service offering against business requirements, to be sure it is fit for its purpose. - ITIL - Key Principles and Models [ITIL Key Principles and Models]

36. A - The Release and Deployment Manager must coordinate tasks, deadlines, and delivery of services and service changes with many different groups within an organization. Of the skills listed, strong project management skills will be most beneficial for this role. - ITIL - Selected Roles [ITIL Selected Roles]

37. D - Big Bang deployments are used for new or modified services that are rolled out to all users at once. A phased approach impacts only a specific group or groups of users at a time. - ITIL - Selected Functions [ITIL Selected Functions]

38. D - Activities that are performed with the intent to address a problem or incident after it has already occurred, such as a server outage, are reactive activities. - ITIL - Selected Functions [ITIL Selected Functions]

39. B - When a service is deemed to no longer meet minimum technical and functional requirements, it is no longer fit for service and should be retired. In order to be renewed, replaced, or refactored, the service must still meet minimum fitness requirements. - ITIL - Key Principles and Models [ITIL Key Principles and Models]

40. C - Service Level Management is found in both the Service Design and Continuous Service Improvement Processes. In Service Design, service levels are designed and negotiated, while in Continuous Service

Improvement, service levels are monitored, evaluated, and improved. - ITIL - Generic Concepts and Definitions [ITIL Generic Concepts and Definitions]

ITIL Foundation
Practice Exam 15
Practice Questions

Test Name:
ITIL Foundation Practice Exam 15
Total Questions: 40
Correct Answers Needed to Pass:
30 (75.00%)
Time Allowed: 60 Minutes

Test Description

This is a cumulative ITIL Foundation test which can be used as a baseline for initial performance. This practice test includes questions from all ITIL question categories.

Test Questions

1. An SLA that states network latency will not exceed 2 milliseconds between 8am and 5pm. Which of the following terms best describe what this SLA is measuring?

   A. Maintainability

   B. Reliability

   C. Resilience

   D. Availability

2. An RFC, a hard drive, an SLA, and an incident report are examples of which of the following?

   A. Resources

   B. Configuration Items

   C. Assets

   D. Business data

3. Which of the following is the measure of average duration between one incident and the next?

   A. Restore Point Objective

   B. Restore Time Objective

   C. Mean Time Between Failure

   D. Mean Time to Restore Service

4. Business investment categories arranged in order of least to most risk are as follows:

   A. Run the business, grow the business, transform the business.

   B. Transform the business, run the business, grow the business

   C. Run the business, transform the business, grow the business

   D. Grow the business, transform the business, run the business

5. Improving an organization's efficiency by providing easily accessible and up to date information is a function of which process?

   A. Knowledge Management

   B. Change Management

   C. Service Asset Management

   D. Configuration Management

6. Why a customer would need a specific service, and subsequently procure that service from a service provider are drivers behind which of the following?

   A. Service Utility

   B. Service Warranty

   C. Service Value

   D. Service Assets

7. The complete set of services offered by a service provider are included in which of the following?

   A. Service Design

   B. Service Portfolio

   C. Service Assets

   D. Service Catalog

8. To which of the following would an organization refer in order to determine if currently offers a service of its own that is comparable to a competitor's service?

   A. Service Catalog

   B. Service Bulletin

   C. Service Warranty

   D. Service Pipeline

9. Define, analyze, approve, and charter are activities undertaken in which process?

   A. Service Portfolio Management

   B. Availability Management

   C. Service Delivery Management

   D. Supplier Management

10. The three components of a service package are:

    A. Service design, service strategy, and service operation

    B. Core service packages, Service strategy packages, and service level agreement packages

    C. Core service package, supporting services package, and service level agreement packages

D. Service warranty, service utility, and service value

11. Monitoring and optimizing the performance of infrastructure components required to support IT services is the responsibility of which ITIL role?

A. Service Desk

B. Service Manager

C. Capacity Manager

D. IT Manager

12. The IT division at True Point Manufacturing has an onsite service and maintenance agreement with a 3rd party company for all of its networking equipment. What is this agreement known as?

A. Service Level Agreement

B. Standard Business Practices

C. Operational Level Agreement

D. Underpinning Contract

13. Authorized copies of software, both custom and commercial-off-the-shelf, is stored in which of the following?

A. Configuration Management Database

B. Service Management Database

C. Definitive Media Library

D. Release Database

14. The network infrastructure at branch office of bank is being upgraded to offer increased bandwidth and speed. In the days after the upgrade, the network team is onsite to provide support in the event of problems or incidents with the new infrastructure. What does ITIL call this type of support?

A. Service Level Agreement

B. Service Management Support

C. Early Life Support

D. Post Release Support

15. Which of the following would be affected by a change in business requirements?

A. Underpinning Contracts

B. All of the above

C. Service Level Agreements

D. Service Metric Thresholds

16. A large accounting company has established a stringent change management process to control the introduction of changes to the company's IT services and their supporting infrastructure. In which Lifecycle Phase is the Change Management process found?

A. Service Operation

B. Service Delivery

C. Service Transition

D. Service Design

17. Attaining market focus and distinguishing capabilities are objectives of which lifecycle phase?

A. Service Transition

B. Service Design

C. Service Strategy

D. Service Operation

18. In which phase of the lifecycle are new and existing services developed and tested?

A. Service Design

B. Service Strategy

C. Service Operation

D. Service Development

19. Service Transition is heavily dependent on technology. Into which two categories does ITIL divide this technology?

A. Configuration Management and Operations Management

B. Network and Server

C. Service Operation and Service Delivery

D. IT Service Management Systems and IT Service Management Technology and Tools

20. According to ITIL, proper assessment of a change will provide the answer to how many questions?

A. 5

B. 10

C. 6

D. Seven

21. John is responsible for ensuring that a specific database report required by the finance team is aligned with that team's business processes, and that the data returned is formatted in a way that is most useful to the finance team. In which of the following roles is John serving?

A. Process Manager

B. Service Owner

C. Service Manager

D. Process Owner

22. Assessment of costs, capabilities, and demand is part of which Service Strategy process?

A. Demand Management

B. Service Portfolio Management

C. Financial Management for IT

D. Budgeting

23. A managed hosting provider needs a more robust network monitoring tool. Which process would be used to gather data to help the organization determine if it is more appropriate to buy a commercial tool or develop one house?

A. FMIT

B. Service Portfolio Management

C. RACI

D. Demand Management

24. Availability, capacity, and continuity are attributes of which of the following?

A. Service Utility

B. Service Warranty

C. Service Strategy

D. ITSM Lifecycle

25. During which lifecycle phase are suppliers categorized in the Supplier and Contract Database (SCD)?

A. Continual Service Improvement

B. Service Design

C. Service Strategy

D. Service Operation

26. In which phase of the lifecycle is the Request Fulfillment process used?

A. Change Management

B. Continual Service Improvement

C. Service Transition

D. Service Operation

27. What is the difference between a Problem and a Known Error?

A. A Known Error has a known root cause and a workaround, while the cause of a Problem is unknown and under investigation.

B. A Known Error has no workaround, while a Problem does.

C. Known Errors are technology related, while Problems are business related.

D. The cause of a Known Error is under investigation, and a Problem is a Known Error that has been resolved.

28. A service catalog is an output of which phase of the service lifecycle?

A. Service Strategy

B. Service Transition

C. Service Design

D. Service Operation

29. Which of the following is not one of the phases of the Service Lifecycle?

A. Service Development

B. Service Transition

C. Service Operation

D. Service Retirement

30. Service Level Agreements (SLAs) provide mutually agreed upon terms for various service performance targets. ITIL categorizes SLAs into which three groups?

A. Service-based, Customer-based, Operations

B. Service-based, Customer-based, and Multi-level

C. Service-based, Customer-based, Transaction-based

D. Service- based, Customer- based, Supplier- based

31. A new managed hosting provider is installing servers, network equipment, and management tools in preparation for offering its services to the public. Which of the following terms best describes the deployment of these systems?

A. Steady state

B. Greenfield

C. Open source

D. Fully managed

32. The database team is testing a new report for release to the legal department. Which of the following will contribute the most to the reliability of the test results?

A. Test data

B. Network infrastructure

C. Storage arrays

D. Bandwidth

33. The Configuration Management System houses which of the following?

A. Supplier and Contact Database

B. All of the above

C. Configuration Management Database

D. Known Error Database

34. Acme Systems uses three service desks located on three different continents to provide round the clock support to their customers. What is this model called?

A. Follow the Sun

B. 24x7x365

C. Multiple Call Center

D. Technical Assistance Center

35. Service investments fall into which three categories?

A. Start the business, run the business, enhance the business

B. Control the business, run the business, grow the business

C. Transform the business, grow the business, and run the business

D. Transform the business, define the business, and run the business

36. Which of these tools and methods is appropriate for informing users of scheduled service changes?

A. Email

B. All of the above

C. Verbal communication

D. Corporate intranet

37. Critical business processes are identified during which activity?

A. Vulnerability Assessment

B. Business Impact Analysis

C. Risk Assessment

D. Threat Assessment

38. A disaster recovery plan that calls for recovery in less than 24 hours requires what type of recovery facility?

A. Hot standby

B. Warm standby

C. Cold standby

D. Reciprocal arrangement

39. An insurance company offers its customers tools to view their policies and statements, as well as pay their bills on the company's website. What are these types of tools called?

A. Self Help

B. Online

C. Portal

D. Web-based

40. Applying context to data results in which of the following?

A. Knowledge

B. Information

C. Facts

D. Wisdom

ITIL Foundation Practice Exam 15 Answer Key and Explanations

1. D - Availability is the ability of a service or component to perform its required function at stated instant or over a stated period of time. - ITIL - Key Principles and Models [ITIL Key Principles and Models]

2. B - A Configuration Item (CI) is any item that supports an IT service. CI's include documents, hardware, software, contracts, and other items necessary for a service offering. - ITIL - Key Principles and Models [ITIL Key Principles and Models]

3. C - Mean Time Between Failure (MTBF) is the average duration between one incident and the next. This is also known as uptime. - ITIL - Generic Concepts and Definitions [ITIL Generic Concepts and Definitions]

4. A - Investments that run the business's current service operations are least risky; investments intended to grow the business are moderately risky; and investments to transform the business are the most risky. - ITIL - Generic Concepts and Definitions [ITIL Generic Concepts and Definitions]

5. A - The goal of the Knowledge Management is to provide accessible, quality, and relevant data to an organization's staff, to improve that organization's efficiency. - ITIL - Selected Processes [ITIL Selected Processes]

6. C - In order for a service to have value, it must provide the functionality a customer requires (Service Utility) and in a guaranteed manner (Service Warranty). - ITIL - Generic Concepts and Definitions [ITIL Generic Concepts and Definitions]

7. B - The Service Portfolio contains the complete set of services managed and offered by a service provider. The Service Portfolio includes the Services Pipeline, the Services Catalogue, and Retired Services. - ITIL - Selected Processes [ITIL Selected Processes]

8. A - The Service Catalog is the complete listing of currently available services a provider is offering. It is one of the three categories of services in the Service Portfolio. - ITIL - Selected Functions [ITIL Selected Functions]

9. A - Service Portfolio Management is an ongoing process an organization uses to define the services it offers, analyze the value of the service portfolio, approve the services to be offered and the resources to support them, and charter the progress of service investments through their lifecycle. - ITIL - Selected Processes [ITIL Selected Processes]

10. C - Service packages are comprised of core services, supporting services, and

service level agreements. These components are often reusable in other service packages. - ITIL - Generic Concepts and Definitions [ITIL Generic Concepts and Definitions]

11. C - The Capacity Manager is responsible for ensuring adequate and appropriate performance and capacity for all IT services. - ITIL - Selected Roles [ITIL Selected Roles]

12. D - An Underpinning Contract (UC) between an organization and an outside supplier supports the organization in the delivery of its services. Third party maintenance coverage of equipment is an example of a UC. - ITIL - Generic Concepts and Definitions [ITIL Generic Concepts and Definitions]

13. C - The Definitive Media Library, or DML, is the official repository of authorized software. This repository also contains documentation and licensing information for the software it houses. - ITIL - Technology and Architecture [ITIL Technology and Architecture]

14. C - Early Life Support (ELS) is intended to offer additional support and assistance immediately after a service deployment, to ensure any issues are ironed out before a deployment is considered complete ELS can include onsite support, dedicated phone support, updated knowledge bases, or service documentation. - ITIL - Key Principles and Models [ITIL Key Principles and Models]

15. B - Changes in business requirements can significantly impact a great many service components, including changes in SLAs, supplier contracts, and metrics used to monitor and measure services. - ITIL - Service Management as a Practice [Service Management as a Practice]

16. C - Service Management is responsible for managing and controlling the introduction of a new or changed service into operation. Configuration management tools and databases are used to track, among other things, what changed, who approved the change, and when it was changed . - ITIL - Selected Processes [ITIL Selected Processes]

17. C - During the Service Strategy phase a service provider, whether an organization or a team within an organization, determines its market focus (where and how it will compete), and distinguishes the capabilities and service assets that will assist the company achieve its focus. - ITIL - Service Lifecycle [The Service Lifecycle]

18. A - During the Service Design phase, an organization designs, develops, and tests new services, as well as modifying and testing existing services. - ITIL - Service Lifecycle [The Service Lifecycle]

19. D - The two types of technology that support Service Transition are IT Service Management Systems, which includes CMDBs and system management tools, and IT Service Management Technology and Tools, which includes service knowledge management tools, data mining tools, and release and deployment technology. - ITIL - Technology and Architecture [ITIL Technology and Architecture]

20. D - There are seven essential questions that must be answered for a change to be appropriately assessed. These are also known as the 7Rs: Who raised the change? What is the reason for the change? What is the return required from the change? What are the risks involved in the change? What resources are required to deliver the change? Who is responsible for the build, test, and implementation of the change? What is the relationship between this change and other changes? - ITIL - Selected Functions [ITIL Selected Functions]

21. D - The Process Owner is responsible for ensuring a service is fit for its intended purpose, and is held accountable for the output of that process - ITIL - Selected Roles [ITIL Selected Roles]

22. C - The goal of the Financial Management for IT (FMIT) process is to clearly define for an organization the costs of providing new and existing services, in order that balance between opportunities and capabilities may be reached. - ITIL - Selected Processes [ITIL Selected Processes]

23. A - Financial Management for IT is used to estimate the cost to deliver a service, data that is subsequently used to develop business models and cost benefit analyses. - ITIL - Selected Processes [ITIL Selected Processes]

24. B - Service warranty provides a customer with reassurance and guarantees that a specific service meets the customer's requirements. Availability, capacity, and continuity are some of the service levels on which service warranty is based. - ITIL - Generic Concepts and Definitions [ITIL Generic Concepts and Definitions]

25. B - During the Service Design process, suppliers required to support an organization's IT services are identified and categorized in the SCD. - ITIL - Selected Processes [ITIL Selected Processes]

26. D - Requests by users for support, documentation, information, or other support for a service are handled by the Service Desk, in the Service Operation phase. - ITIL - Selected Processes [ITIL Selected Processes]

27. A - The root cause of a Known Error has been determined and a workaround has been identified. Problems do not have a known root cause and are under investigation. - ITIL - Generic Concepts and

Definitions [ITIL Generic Concepts and Definitions]

28. C - During the Service Design phase, new services are developed and existing services modified. Outputs of this phase include service catalogs, service level agreements, and business requirements. - ITIL - Service Lifecycle [The Service Lifecycle]

29. D - The Service Lifecycle is composed of 5 interdependent phases: Service Strategy, Service Design, Service Transition, Service Operation, and Continual Service Improvement. - ITIL - Service Lifecycle [The Service Lifecycle]

30. B - ITIL categorizes SLAs as either Service-based, Customer-based, or Multi-level. - ITIL - Key Principles and Models [ITIL Key Principles and Models]

31. B - A greenfield deployment is the implementation of systems, services, tools, or technology where there had previously been none. Only a new service will have a greenfield deployment; changes to existing services are considered modifications. - ITIL - Technology and Architecture [ITIL Technology and Architecture]

32. A - Good solid test data is essential conduct a test with reliable results. Test data should mimic production data as closely as possible, while following company policies regarding data confidentiality and security. - ITIL - Technology and Architecture [ITIL Technology and Architecture]

33. B - All of the above. The Configuration Management System contains all the tools and databases to manage an IT service provider's configuration data, as well as data on known errors, suppliers, business users, and customers. - ITIL - Key Principles and Models [ITIL Key Principles and Models]

34. A - The Follow the Sun model uses multiple call centers in multiple time zones to provide round the clock support for customers. - ITIL - Key Principles and Models [ITIL Key Principles and Models]

35. C - Organizational service investment are divided into 3 categories. Transform the business (TTB) investments move the organization into new market areas. Grow the business investments (GTB) enable an organization to increase the scope of its service offerings. Run the business (RTB) investments are intended to maintain the current service offerings. - ITIL - Selected Processes [ITIL Selected Processes]

36. B - Any organizationally appropriate communication method can be used to communicate service changes to users. For some organizations this might be an all hands staff meeting, while other organizations will use an electronic method such as email or text messaging. - ITIL - Technology and Architecture [ITIL Technology and Architecture]

37. B - A Business Impact Analysis (BIA) identifies critical business processes, as well as the possible damage or loss caused by disruption to those processes. - ITIL - Selected Processes [ITIL Selected Processes]

38. A - A hot standby facility permits an organization to recover its critical business functions in less than 24 hours, and typically much more quickly than this. - ITIL - Selected Processes [ITIL Selected Processes]

39. A - Self Help tools provide customers with easy access to information and services without the assistance of the organization's staff. Viewing statements and documentation, paying bills, and password resets are examples of support requests that can be serviced with Self Help tools - ITIL - Technology and Architecture [ITIL Technology and Architecture]

40. B - Information is derived from applying context, or meaning, to data. - ITIL - Key Principles and Models [ITIL Key Principles and Models]

Additional Resources

Exam Taking Tips

Studying for a multiple choice exam entails preparing in a unique way as opposed to other types of tests. The ITIL Foundation exam asks one to recognize correct answers among a set of four options. The extra options that are not the correct answer are called the "distracters"; and their purpose, unsurprisingly, is to distract the test taker from the actual correct answer among the bunch.

Students usually consider multiple choice exams as much easier than other types of exams; this is not necessarily true with the ITIL Foundation exam. Among these reasons are:

- Most multiple choice exams ask for simple, factual information; unlike the Foundation exam which often requires the student to apply knowledge and make a best judgment.
- The majority of multiple choice exams involve a large quantity of different questions – so even if you get a few incorrect, it's still okay. The Foundation exam covers a broad set of material, often times in greater depth than other certification exams.

Regardless of whether or not multiple choice testing is more forgiving; in reality, one must study immensely because of the sheer volume of information that is covered.

Although 60 minutes may seem like more than enough time for a multiple choice exam of 40 questions; time management remains a crucial factor in succeeding and doing well. You should always try and answer all of the questions you are confident about first, and then go back to those items you are not sure about afterwards. Always read *carefully* through the entire test as well, and do your best to not leave any question blank upon submission– even if you do not readily know the answer.

Many people do very well with reading through each question and not looking at the options before trying to answer. This way, they can steer clear (usually) of being fooled by one of the "distracter" options or get into a tug-of-war between two choices that both have a good chance of being the actual answer.

Never assume that "all of the above" or "none of the above" answers are the actual choice. Many times they are, but in recent years they have been used much more frequently as distracter options on standardized tests. Typically this is done in an effort to get people to stop believing the myth that they are always the correct answer.

You should be careful of negative answers as well. These answers contain words such as "none", "not", "neither", and the like. Despite often times being very confusing, if you read these types of questions and answers carefully, then you should be able to piece together which is the correct answer. Just take your time!

Never try to overanalyze a question, or try and think about how the test givers are trying to lead astray potential test takers. Keep it simple and stay with what you know.

If you ever narrow down a question to two possible answers, then try and slow down your thinking and think about how the two different options/answers differ. Look at the question again and try to apply how this difference between the two potential answers relates to the question. If you are convinced there is literally no difference between the two potential answers (you'll more than likely be wrong in assuming this), then take another look at the answers that you've already eliminated. Perhaps one of them is actually the correct one and you'd made a previously unforeseen mistake.

On occasion, over-generalizations are used within response options to mislead test takers. To help guard against this, always be wary of responses/answers that use absolute words like "always", or "never". These are less likely to actually be the answer than phrases like "probably" or "usually" are. Funny or witty responses are also, most of the time, incorrect – so steer clear of those as much as possible.

Although you should always take each question individually, "none of the above" answers are usually less likely to be the correct selection than "all of the above" is. Keep this in mind with the understanding that it is not an absolute rule, and should be analyzed on a case-by-case (or "question-by-question") basis.

Looking for grammatical errors can also be a huge clue. If the stem ends with an indefinite article such as "an" then you'll probably do well to look for an answer that begins with a vowel instead of a consonant. Also, the longest response is also oftentimes the correct one, since whoever wrote the question item may have tended to load the answer with qualifying adjectives or phrases in an effort to make it correct. Again though, always deal with these on a question-by-question basis, because you could very easily be getting a question where this does not apply.

Verbal associations are oftentimes critical because a response may repeat a key word that was in the question. Always be on the alert for this. Playing the old Sesame Street game "Which of these things is not like the other" is also a very solid strategy, if a bit preschool. Sometimes many of a question's distracters will be very similar to try to trick you into thinking that one choice is related to the other. The answer very well could be completely unrelated however, so stay alert.

Just because you have finished a practice test, be aware that you are not done working. After you have graded your test with all of the necessary corrections, review it and try to recognize what happened in the answers that you got wrong. Did you simply not know the qualifying correct information? Perhaps you were led astray by a solid distracter answer? Going back through your corrected test will give you a leg up on your next one by revealing your tendencies as to what you may be vulnerable with, in terms of multiple choice tests.

It may be a lot of extra work, but in the long run, going through your corrected multiple choice tests will work wonders for you in preparation for the real exam. See if you perhaps misread the question or even missed it because you were unprepared. Think of it like instant replays in professional sports. You are going back and looking at what you did on the big stage in the past so you can help fix and remedy any errors that could pose problems for you on the real exam.